Incorporating Knowledge Sources into Statistical Speech Recognition

Lecture Notes in Electrical Engineering

Incorporating Knowledge Sources into Statistical Speech Recognition
Sakti, Sakriani, Markov, Konstantin, Nakamura, Satoshi, and Minker, Wolfgang
978-0-387-85829-6

Intelligent Technical Systems
Martínez Madrid, Natividad; Seepold, Ralf E.D. (Eds.)
978-1-4020-9822-2

Languages for Embedded Systems and their Applications
Radetzki, Martin (Ed.)
978-1-4020-9713-3

Multisensor Fusion and Integration for Intelligent Systems
Lee, Sukhan; Ko, Hanseok; Hahn, Hernsoo (Eds.)
978-3-540-89858-0

Designing Reliable and Efficient Networks on Chips
Murali, Srinivasan
978-1-4020-9756-0

Trends in Communication Technologies and Engineering Science
Ao, Sio-Iong; Huang, Xu; Wai, Ping-kong Alexander (Eds.)
978-1-4020-9492-7

Functional Design Errors in Digital Circuits: Diagnosis Correction and Repair
Chang, Kai-hui, Markov, Igor, Bertacco, Valeria
978-1-4020-9364-7

Traffic and QoS Management in Wireless Multimedia Networks: COST 290 Final Report
Koucheryavy, Y., Giambene, G., Staehle, D., Barcelo-Arroyo, F., Braun, T., Siris, V. (Eds.)
978-0-387-85572-1

Proceedings of the 3rd European Conference on Computer Network Defense
Siris, V.; Ioannidis, S.; Anagnostakis, K.; Trimintzios, P. (Eds.)
978-0-387-85554-7

Intelligentized Methodology for Arc Welding Dynamical Processes: Visual Information Acquiring, Knowledge Modeling and Intelligent Control
Chen, Shan-Ben, Wu, Jing
978-3-540-85641-2

Proceedings of the European Computing Conference: Volume 2
Mastorakis, Nikos, Mladenov, Valeri, Kontargyri, Vassiliki T. (Eds.)
978-0-387-84818-1

Proceedings of the European Computing Conference: Volume 1
Mastorakis, Nikos, Mladenov, Valeri, Kontargyri, Vassiliki T. (Eds.)
978-0-387-84813-6

Electronics System Design Techniques for Safety Critical Applications
Sterpone, Luca
978-1-4020-8978-7

Data Mining and Applications in Genomics
Ao, Sio-Iong
978-1-4020-8974-9

Continued after index

Sakriani Sakti • Konstantin Markov •
Satoshi Nakamura • Wolfgang Minker

Incorporating Knowledge Sources into Statistical Speech Recognition

Sakriani Sakti
NICT/ATR Spoken Language
Communication Research Laboratories
Keihanna Science City
Kyoto, Japan

Konstantin Markov
NICT/ATR Spoken Language
Communication Research Laboratories
Keihanna Science City
Kyoto, Japan

Satoshi Nakamura
NICT / ATR Spoken Language
Communication Research Laboratories
Keihanna Science City
Kyoto, Japan

Wolfgang Minker
University of Ulm
Ulm, Germany

ISBN 978-0-387-85829-6 e-ISBN 978-0-387-85830-2
DOI: 10.1007/978-0-387-85830-2

Library of Congress Control Number: 2008942803

Printed on acid-free paper.

springer.com

This book is dedicated
to our parents and families
for their support and endless love

Preface

State-of-the-art automatic speech recognition (ASR) systems use statistical data-driven methods based on hidden Markov models (HMMs). Although such approaches have proved to be efficient choices, ASR systems often perform much worse than human listeners, especially in the presence of unexpected acoustic variability. To improve performance, we usually rely on collecting more data to train more detailed models. However, such resources are rarely available, since the presence of variabilities in speech arise from many different factors, and thus a huge amount of training data is required to cover all possible variabilities. In other words, it is not enough to handle these variabilities by relying solely on statistical models. The systems need additional knowledge on speech that could help to handle these sources of variability. Otherwise, only a limited level of success could be achieved.

Many researchers are aware of this problem, and thus various attempts to integrate more explicitly knowledge-based and statistical approaches have been made. However, incorporating various additional knowledge sources often leads to a complicated model, where achieving optimal performance is not feasible due to insufficient resources or data sparseness. As a result, input space resolution may be lost due to non-robust estimates and the increased number of unseen patterns. Moreover, decoding with large models may also become cumbersome and sometimes even impossible.

This book addresses the problem of developing efficient ASR systems that can maintain a balance between utilizing wide-ranging knowledge of speech variability while keeping the training/recognition effort feasible, of course while also improving speech recognition performance. In this book, an efficient general framework to incorporate additional knowledge sources into state-of-the-art statistical ASR systems is provided. It can be applied to many existing ASR problems with their respective model-based likelihood functions in flexible ways.

Since there are various types of knowledge sources from different domains, it may be difficult to formulate a probabilistic model without learning the dependencies between the sources. To solve such problems in a unified way, the work reported in this book adopts the Bayesian network (BN) framework. This approach allows the probabilistic relationship between information sources to be learned. Another advantage of the BN framework lies in the fact that it facilitates the decomposition of the joint probability density function (PDF) into a linked set of local conditional PDFs based on the junction tree algorithm. Consequently, a simplified form of the model can be constructed and reliably estimated using a limited amount of training data.

This book focuses on the acoustic modeling problem as arguably the central part of any speech recognition system. The incorporation of various knowledge sources, including background noises, accent, gender and wide phonetic knowledge information, in modeling is also discusses. Such an application often suffers from a sparseness of data and memory constraints. First, the additional sources of knowledge are incorporated at the HMM state distribution. Then, these additional sources of knowledge are incorporated at the HMM phonetic modeling. The presented approaches are experimentally verified in the large-vocabulary continuous-speech recognition (LVCSR) task. The book closes with a summary of the described methods and the results of the evaluations.

Contents

1 Introduction and Book Overview ... 1
1.1 Automatic Speech Recognition - A Way of Human-Machine Communication ... 1
1.2 Approaches to Speech Recognition ... 4
1.2.1 Knowledge-based Approaches ... 4
1.2.2 Corpus-based Approaches ... 6
1.3 State-of-the-art ASR Performance ... 7
1.4 Studies on Incorporating Knowledge Sources ... 10
1.4.1 Sources of Variability in Speech ... 10
1.4.2 Existing Ways of Incorporating Knowledge Sources ... 12
1.4.3 Major Challenges to Overcome ... 15
1.5 Book Outline ... 16

2 Statistical Speech Recognition ... 19
2.1 Pattern Recognition Overview ... 19
2.2 Theory of Hidden Markov Models ... 22
2.2.1 Markov Chain ... 22
2.2.2 General form of an HMM ... 23
2.2.3 Principle Cases of HMM ... 25
2.3 Pattern Recognition for HMM-Based ASR Systems ... 35
2.3.1 Front-end Feature Extraction ... 36
2.3.2 HMM-Based Acoustic Model ... 43
2.3.3 Pronunciation Lexicon ... 49
2.3.4 Language Model ... 50
2.3.5 Search Algorithm ... 51

3 Graphical Framework to Incorporate Knowledge Sources ... 55
3.1 Graphical Model Representation ... 56
3.1.1 Probability Theory ... 56
3.1.2 Graphical Model ... 59
3.1.3 Junction Tree Algorithm ... 63

3.2 Procedure of GFIKS 68
3.2.1 Causal Relationship between Information Sources 70
3.2.2 Direct Inference on Bayesian Network 71
3.2.3 Junction Tree Decomposition 72
3.2.4 Junction Tree Inference 75
3.3 Practical Issues of GFIKS 75
3.3.1 Types of Knowledge Sources 75
3.3.2 Different Levels of Incorporation 76

4 Speech Recognition Using GFIKS 79
4.1 Applying GFIKS at the HMM State Level 79
4.1.1 Causal Relationship Between Information Sources 80
4.1.2 Inference 81
4.1.3 Enhancing Model Reliability 81
4.1.4 Training and Recognition Issues 82
4.2 Applying GFIKS at the HMM Phonetic-unit Level 83
4.2.1 Causal Relationship between Information Sources 83
4.2.2 Inference 85
4.2.3 Enhancing the Model Reliability 85
4.2.4 Deleted Interpolation 86
4.2.5 Training and Recognition Issues 86
4.3 Experiments with Various Knowledge Sources 87
4.3.1 Incorporating Knowledge at the HMM State Level 87
4.3.2 Incorporating Knowledge at the HMM Phonetic-unit Level 116
4.4 Experiments Summary and Discussion 132

5 Conclusions and Future Directions 139
5.1 Conclusions 139
5.1.1 Theoretical Issues 139
5.1.2 Application Issues 140
5.1.3 Experimental Issues 141
5.2 Future Directions: A Roadmap to a Spoken Language Dialog System 142

A Speech Materials 145
A.1 AURORA TIDigit Corpus 145
A.2 TIMIT Acoustic-Phonetic Speech Corpus 146
A.3 Wall Street Journal Corpus 148
A.4 ATR Basic Travel Expression Corpus 150
A.5 ATR English Database Corpus 150

B ATR Software Tools 153
B.1 Generic Properties of ATRASR 153
B.2 Data Preparation 153
B.3 SSS Data Generating Tools 155
B.4 Acoustic Model Training Tools 155
B.5 Language Model Training Tools 157
B.6 Recognition Tools 157

C Composition of Bayesian Wide-phonetic Context 163
C.1 Proof using Bayes's Rule 163
C.2 Variants of Bayesian Wide-phonetic Context Model 164

D Statistical Significance Testing 169
D.1 Statistical Hypothesis Testing 169
D.2 The Use of the Sign Test for ASR 172

References 175

Index 189

List of Figures

1.1 A machine that recognizes the speech waveform of a human utterances as "Good night." . . . 2
1.2 Knowledge-based ASR system. . . . 4
1.3 Speech spectrogram reading, which corresponds to the word sequence of "Good night" . . . 5
1.4 Corpus-based statistical ASR system. . . . 6
1.5 2003 NIST's benchmark ASR test history (After Pallett, 2003, ©2003 IEEE). . . . 7
1.6 TC-STAR ASR evaluation campaign (After Choukri, 2007, ©TC-STAR). . . . 8
1.7 "S" curve of ASR technology progress and the predicted performance from combining deep knowledge with a statistical approach. . . . 9
1.8 Incorporating knowledge into a corpus-based statistical ASR system. . . . 16

2.1 Pattern recognition: Establishing mapping from multi-dimensional measurement space X to three-class target decision space Y. . . . 20
2.2 Pattern recognition approach for ASR: Establish mapping from measurement space X of speech signal to target decision space Y of word strings. . . . 22
2.3 Simple three-state Markov chain for daily weather. . . . 22
2.4 HMM of the daily weather, where there is no deterministic meaning on any state. . . . 24
2.5 Left-to-right HMM of the daily weather. . . . 25
2.6 Process flow on trellis diagram of 3-state HMM with time length T. . . . 26
2.7 Forward probability function representation (for j=1). . . . 27
2.8 Backward probability function representation (for i=1). . . . 28

2.9 Example of finding the best path on a trellis diagram using the Viterbi algorithm. 30
2.10 Graphical interpretation of the EM algorithm. 31
2.11 Forward-backward probability function representation. 32
2.12 A generic automatic speech recognition system, composed of five components: feature extraction, acoustic model, pronunciation lexicon, language model and search algorithm. . . . 36
2.13 Source-Filter model of the speech signal $x[n] = e[n] * h[n]$. 37
2.14 Source-filter separation by cepstral analysis. 37
2.15 (a) A windowed speech waveform. (b) The spectrum of Figure 2.15(a). (c) The resulting cepstrum. (d) The Fourier transform of the low-quefrency component. 38
2.16 MFCC feature extraction technique, which generates a 25-dimensional feature vector x_t for each frame. 41
2.17 A summary of feature extraction process, producing a feature vector which correspond to one point in a multi-dimensional space. 43
2.18 Discrete HMM observation density where the emission statistics or HMM state output probabilities are represented by discrete symbols. 44
2.19 Continuous GMM, where the continuous observation space is modeled using mixture Gaussians (state-specific). They are weighted and added to compute the emission statistic likelihoods (HMM state output probabilities). 45
2.20 Structure example of the monophone $/a/$ HMM acoustic model. 46
2.21 Structure example of the triphone $/a^-, a, a^+/$ HMM acoustic model. 47
2.22 Shared-state structures of the triphone $/a^-, a, a^+/$ HMM acoustic model. 47
2.23 An example of a phonetic decision tree for HMM state of the triphone with the central phoneme $/ay/$. 48
2.24 Contextual splitting and temporal splitting of SSS algorithm (After Jitsuhiro, 2005). 49
2.25 Example of a tree-based pronunciation lexicon. 50
2.26 Multi-level probability estimation of statistical ASR. 52

3.1 Incorporating knowledge into corpus-based statistical ASR system. 55
3.2 Two equivalent models that can be obtained from each other through arc reversal of Bayes's rule, since P(a,b)=P(b,a). 60
3.3 Graphical representation of $P(a|b_1, b_2, ..., b_n)$. 60
3.4 Three BNs with different arrow directions over the same random variables a, b, and c. They appear in the case of serial, diverging, and converging connection, respectively. 61

3.5 Example of BN topology describing conditional relationship among a, b, c, d, e, f, g and h. 63
3.6 Moral and triangulated graph of Figure 3.5. 64
3.7 Junction graph of Figure 3.5. 66
3.8 The resulting junction tree. 66
3.9 Clique $C_1 = [a, b, d]$ in the original graph of Figure 3.5. 67
3.10 General procedure of GFIKS (graphical framework to incorporate additional knowledge sources). 69
3.11 (a) BN topology describing the conditional relationship between data D and model M. (b) BN topology describing the conditional relationship among D, M, and additional knowledge K. 70
3.12 Examples of BN topologies describing the conditional relationship among data D, model M, and several knowledge sources $K_1, K_2, ..., K_N$. 71
3.13 (a) BN topology describing conditional relationship among D, M, K_1, and K_2. (b) Moral and triangulated graph of Figure 3.13(a). (c) Equivalent BN topology. (d) Moral and triangulated graph of Figure 3.13(c). (e) Junction tree of Figure 3.13(d). 73
3.14 (a) Equivalent BN topology of the BN shown in Figure 3.12(a). (b) Corresponding junction tree. 74
3.15 Incorporating knowledge sources into HMM state (denoted by a small box) and phonetic unit level (denoted by a large box). . 77

4.1 (a) Applying GFIKS at the HMM state level. (b) BN topology structure describing the conditional relationship between HMM state Q and observation vector X. 80
4.2 BN topology structure after incorporating additional knowledge sources $K_1, K_2, ..., K_N$ in HMM state distribution $P(X, Q)$ (assuming that all $K_1, K_2, ..., K_N$ are independent given Q). 81
4.3 Example of observation space modeling by BN, where each value of K_i corresponds to a different Gaussian. 82
4.4 (a) Applying GFIKS at the HMM phonetic-unit level. (b) BN topology structure describing the conditional relationship between HMM phonetic model λ and observation segment X_s. . 84
4.5 BN topology structure after incorporating additional knowledge sources $K_1, K_2, ..., K_N$ in HMM phonetic model $P(X_s, \lambda)$ (assuming that all $K_1, K_2, ..., K_N$ are independent given λ). 84
4.6 Rescoring procedure with the composition models. 87
4.7 BN topology structure showing the conditional relationship among HMM state Q, observation vector X, and additional knowledge source of gender information G. 88

4.8 Recognition accuracy rates of proposed HMM/BN, which are comparable with those of other systems from the "Hub and Spoke Paradigm for Continuous Speech Recognition Evaluation" for primary condition of WSJ Hub2-5k task. 93
4.9 BN topology structure describing the conditional relationship between HMM state Q, observation vector X, and additional knowledge sources of noise type N and SNR value S. 94
4.10 Comparison of different systems: HMM, DBN (Bilmes et al., 2001), and proposed HMM/BN 98
4.11 BN topologies of the left state (a), center state (b), and right state (c) of LR-HMM/BN for modeling a pentaphone context $/a^{--}, a^{-}, a, a^{+}, a^{++}/$.................................. 99
4.12 BN topologies of the left state (a), center state (b), and right state (c) of LRC-HMM/BN, for modeling a pentaphone context $/a^{--}, a^{-}, a, a^{+}, a^{++}/$.............................. 100
4.13 Observation space modeling by BN, where a different value of second following context C_R corresponds to a different Gaussian. 101
4.14 Knowledge-based phoneme classes of the observation space..... 102
4.15 Determining distance metric by Euclidean distance............ 103
4.16 Data-driven phoneme classes of observation space............. 103
4.17 Recognition accuracy rates of pentaphone LR-HMM/BN using knowledge-based second preceding and following context clustering.. 106
4.18 Recognition accuracy rates of pentaphone LRC-HMM/BN using knowledge-based second preceding and following context clustering.. 107
4.19 Recognition accuracy rates of pentaphone LR-HMM/BN and LRC-HMM/BN using data-driven Gaussian clustering......... 108
4.20 Comparing recognition accuracy rates of triphone HMM and pentaphone HMM/BN models with a fixed and a varied number of mixture components per state, but having the same 15 mixture components per state on average. 109
4.21 Topology of fLRC-HMM/BN for modeling a pentaphone context $/a^{--}, a^{-}, a, a^{+}, a^{++}/$, where state PDF has additional variables C_L and C_R representing the second preceding and following contexts, respectively. 110
4.22 (a) fLRCG-HMM/BN topology with additional knowledge G, C_L and C_R, (b) fLRCA-HMM/BN topology with additional variables A, C_L, and C_R, and (c) fLRCAG-HMM/BN topology with additional knowledge A, G, C_L, and C_R................. 111
4.23 Recognition accuracy rates of proposed HMM/BN models having identical numbers of 5, 10, and 20 mixture components per state.. 113

4.24 Comparing recognition accuracy rates of different systems: triphone HMM baseline, pentaphone HMM baseline, and the proposed pentaphone HMM/BN models having the same five mixture components per state. 115
4.25 BN topology structure describing the conditional relationship among X_s, λ, C_L, and C_R. 116
4.26 (a) Equivalent BN topology. (b) Moral and triangulated graph of Figure 4.26(a). (c) Junction tree of Figure 4.26(b). 117
4.27 (a) Conventional triphone model, (b) Conventional pentaphone model, (c) Bayesian pentaphone model composition C1L3R3, consisting of the preceding/following triphone-context unit and center-monophone unit. 119
4.28 Rescoring procedure with pentaphone composition models: C1L3R3 or C3L4R4. 120
4.29 N-best rescoring mechanism. 121
4.30 Recognition accuracy rates of Bayesian triphone model. 122
4.31 Recognition accuracy rates of Bayesian pentaphone models. 124
4.32 Relative reductions in WER by Bayesian triphone C1L2R2 model from monophone baseline and by Bayesian pentaphone C1L3R3 model from triphone baseline. 125
4.33 Recognition accuracy rates of conventional pentaphone C5 and proposed Bayesian pentaphone C1L3R3 models with different amounts of training data. 126
4.34 BN topology structure describing the conditional relationship among X_s, λ, C_L, C_R, A, and G. 127
4.35 (a) Equivalent BN topology of Figure 4.34. (b) Moral and triangulated graph of Figure 4.35(a). (c) Corresponding junction tree. 127
4.36 Rescoring procedure with the accent-gender-dependent pentaphone composition models: C1L3R3, C1L3R3-A, C1L3R3-G, and C1L3R3-AG. 128
4.37 Comparing recognition accuracy rates of different systems triphone HMM baseline, pentaphone HMM baseline, and proposed pentaphone models having the same 5, 10, and 20 mixture components per state. 130
4.38 Comparing recognition accuracy rates of different systems: triphone HMM baseline, pentaphone HMM baseline, and proposed models incorporating knowledge sources at HMM state and phonetic unit levels. 136

5.1 Roadmap to spoken language dialog system incorporating other knowledge sources at higher ASR levels. 143

B.1 The ATRASR phoneme-based SSS data creation for phone-unit model training. 156

B.2 The ATRASR topology training for each phone acoustic-unit model. 158
B.3 The ATRASR embedded training for a whole HMnet. 159
B.4 The recognition process using ATRASR tools. 160

C.1 Bayesian pentaphone model composition. (a) is C5, the conventional pentaphone model, (b) is Bayesian C1L3R3, which is composed of the preceding/following triphone-context unit and center-monophone unit, (c) is Bayesian C3L4R4, which is composed of the preceding/following tetraphone-context unit and center-triphone-context unit, (d) is Bayesian C1Lsk3Rsk3, which is composed of the preceding/following skip-triphone-context unit and center-monophone unit, and (e) is Bayesian C1C3Csk3, which is composed of the center skip-triphone-context unit, center triphone-context unit and center-monophone unit. 167

D.1 The distribution of population according to the null hypothesis (H_0 is true), with upper-tail of rejection region for $P \leq \alpha$. 171

List of Tables

4.1 *English phoneme set.* 90
4.2 *1993 Hub and Spoke CSR evaluation on Hub 2: 5k read WSJ task (Kubala et al., 1994; Pallett et al., 1994).* 92
4.3 *HMM/BN system performance on Hub 2: 5k read WSJ task.* 93
4.4 *Recognition accuracy rates (%) for proposed HMM/BN on AURORA2 task.* 97
4.5 *Knowledge-based phoneme classes based on manner of articulation.* 101
4.6 *Recognition accuracy rates (%) for proposed pentaphone HMM/BN model using fLRC-HMM/BN (see Figure 4.22) on a test set of matching accents with different numbers of mixture components.* 114
4.7 *Recognition accuracy rates (%) for proposed pentaphone HMM/BN model using fLRC-HMM/BN (see Figure 4.22) on a test set of mismatched accents with 15 mixture components.* 115
4.8 *Recognition accuracy rates (%) for proposed Bayesian pentaphone C1L3R3-AG (see Eq. (4.30)) on a test set of matching accents with different numbers of mixture components.* 131
4.9 *Recognition accuracy rates (%) for proposed Bayesian pentaphone C1L3R3-AG model (see Eq. (4.30)) on a test set of mismatched accents with 15 mixture components.* 132
4.10 *Summary of incorporating various knowledge sources at the HMM state level.* 134
4.11 *Summary of incorporating various knowledge sources at the HMM phonetic unit level.* 135

A.1 *Dialect distribution of speakers.* 147
A.2 *Speech materials of TIMIT database.* 148
A.3 *Statistics on the TIMIT database.* 148
A.4 *Text sentence materials of ATR English speech database.* 151
A.5 *Speech materials of ATR English speech database.* 151

Glossary

AM	Acoustic model
ARPA	Advanced Research Projects Agency
ASR	Automatic speech recognition
A-STAR	Asian speech translation advanced research
ATR	Advanced Telecommunication Research
AUS	Australian
BN	Bayesian network
BRT	British
BTEC	Basic travel expression corpus
BU	Boston University
C1	Center monophone unit
C3	Center triphone context
Csk3	Center skip-triphone context
C5	Center pentaphone context
CCCC	CSR corpus coordinating committee
CNRS-LIMSI	France's National Center for Scientific Research
CPD	Conditional probability distribution
CPT	Conditional probability table
CSR	Continuous speech recognition
C-STAR	Consortium for speech translation advanced research
CU	Cambridge University
DAG	Directed acyclic graph
DARPA	Defense Advanced Research Projects Agency
DBN	Dynamic Bayesian network
DCT	Discrete cosine transform
DEL	Deletions
DI	Deleted interpolation
DSR	Distributed speech recognition
EDB	English database
ELRA	European language resources association

EM	Expectation-maximization
EPPS	European Parliament Plenary Sessions
fLRC-HMM/BN	Full HMM/BN for left, right and center state
fLRCA-HMM/BN	Full HMM/BN for left, right and center state, including accent dependency
fLRCAG-HMM/BN	Full HMM/BN for left, right and center state, including accent and gender dependency
fLRG-HMM/BN	Full HMM/BN for left, right and center state, including gender dependency
FFT	Fast Fourier transform
GDHMM	Gender-dependent Hidden Markov model
GFIKS	Graphical framework to incorporate additional knowledge sources
GIHMM	Gender-independent Hidden Markov model
GMM	Gaussian mixture model
HMM	Hidden Markov model
ICASSP	International conference on acoustics, speech and signal processing
ICSI	International Computer Science Institute
ICSLP	International conference on spoken language processing
IEEE	Institute of Electrical and Electronics Engineers
IEICE	Institute of Electronics, Information and Communication Engineers
Imp	Improvement
INS	Insertions
L3	Left triphone context
L4	Left tetraphone context
LM	Language model
LPC	Linear prediction coefficients
LRC-HMM/BN	HMM/BN for left, right and center state
LR-HMM/BN	HMM/BN for left and right state
Lsk3	Left skip-triphone context
LVCSR	Large-vocabulary continuous-speech recognition
MAD	Machine translation aided dialogue
MAP	Maximum *a posteriori*
MDL	Minimum description length
MFCC	Mel-frequency cepstral coefficients
MIT	Massachusetts Institute of Technology
ML	Maximum likelihood
MLLR	Maximum likelihood linear regression
MSG	Modulation-filtered spectrogram
MT	Machine translation
NIST	National Institute of Standards and Technology
NOVO	Noise voice composition
PDF	Probability density function

PLP	Perceptual linear prediction
PMC	Parallel model combination
R3	Right triphone context
R4	Right tetraphone context
Rel	Relative
Resc	Rescoring
Rsk3	Right skip-triphone context
S2ST	Speech-to-speech translation
SD	Speaker dependent
SI	Speaker independent
SIL	Silence
SLC	Spoken Language Communication
SNR	Signal-to-noise ratio
SSS	Successive state splitting
STQ	Speech processing, transmission and quality
SUB	Substitutions
SWB	Switchboard
TC-STAR	Technology and corpora for speech to speech translation research
TI	Texas Instrument
US	United States
VQ	Vector quantization
WER	Word error rate
WFST	Weighted finite state transducers
WSJ	Wall Street journal

1

Introduction and Book Overview

1.1 Automatic Speech Recognition - A Way of Human-Machine Communication

The continuous growth of information technology is having an increasingly large impact on many aspects of our daily lives. The issue of communication via speech between human beings and information-processing machines is also becoming more important (Holmes and Holmes, 2001).

A common dream is to realize a technology that allows humans to communicate or have dialogs with machines through natural and spontaneous speech, since in many cases of human-machine communication, speech is the most convenient way. It is the most natural modality for humans and thus requires no special user training (Lea, 1986). As a communication channel for human expression, speech also provides the highest capacity. This has been quantitatively demonstrated by Pierce and Kerlin (1957) and also Turn (1974), where spontaneous speech was shown to have a typical transmission rate of around 2.0 to 3.6 words per second; in contrast, handwriting conveys only about 0.4 words per second, and typing, by a skilled typist, achieves about 1.6 to 2.5 words per second. Speech communication also offers obvious benefits for individuals challenged with a variety of physical disabilities, such as loss of sight or limitations in physical motion and motor skills (Lea, 1986).

A fundamental technology for achieving a speech-oriented interface is development of automatic speech recognition (ASR): A machine that can automatically recognize naturally spoken words uttered by humans. A speech waveform is produced by a sound source that propagates though the vocal tract (from larynx to lips) with different resonance properties (e.g. different formant frequencies for different vowel sound). Designing an ASR system mostly involves dealing with the acoustic properties of speech sounds and their relationship to the basic sounds of a human language including phonemes, words, phrases, and sentences (Juang and Rabiner, 2005). Figure 1.1 shows an example of a machine that recognizes the speech waveform of a human utterance as "Good night."

S. Sakti et al., *Incorporating Knowledge Sources into Statistical Speech Recognition*,
Lecture Notes in Electrical Engineering 42, DOI: 10.1007/978-0-387-85830-2_1,

Fig. 1.1. A machine that recognizes the speech waveform of a human utterances as "Good night."

Researchers have been working in speech recognition technology for many decades. Let's take a brief look at some of the major milestones in the history of this work:

- In the 1920s: single-word recognition

  The first machine to recognize speech to any significant degree may have been a commercial dog toy named "Rex," which was manufactured in the 1920s. As described in a review paper by Davis and Selfridge (1962), this canine machine would respond only to acoustic energy around 500 Hz, which is expressed in the vowel sound of the dog's single-word name "Rex." However, this simple machine was unable to reject many other words or non-speech sounds that produced sufficient 500-Hz energy (Gold and Morgan, 1999).

- In the 1950s: Single-speaker, isolated-digit (order of 10 words) recognition

  Many works related to speech analysis were also carried out in the 1930s and 1940s, but the next complete system of any significance was isolated-digit recognition (order of 10 words) for a single speaker, developed at Bell Labs in the early 1950s and published in (Davis et al., 1952). Juang and Rabiner (2005) explained how it used the formant frequency estimated during the vowel regions of each digit, and this may have been the first true word recognizer.

- In the 1960s: Multiple-speaker, small-vocabulary (order of 10-100 words) isolated-word recognition

  In the 1960s, digit recognizers become better, achieving good accuracy for multiple speakers (Juang and Rabiner, 2005). Many ASR systems were able to recognize small vocabularies (order of 10-100 words) of isolated words, based on simple acoustic-phonetic properties of speech sounds (Hughes, 1959; Denes and Mathews, 1960; Petrick and Willet, 1960).

- In the 1970s: Multiple-speaker, medium-vocabulary (order of 100-1000 words) continuous-speech recognition

  Speech recognition research in the 1970s achieved a number of significant milestones, and several ASR systems were able to recognize medium-sized vocabularies (order of 100-1000 words) for multiple speakers (Lesser et al., 1975; Lowerre, 1976; Wolf and Woods, 1977). This period began to focus on more difficult continuous speech recognition tasks, where speech input was uttered in its normal, connected form (Newell et al., 1971). The key technologies based on codification of human knowledge, typically in the form of rules, become widely used in a number of disciplines.

- In the 1980s: Multiple-speaker, large-vocabulary (order of 1000-10,000 words) continuous-speech recognition

  Researchers started to tackle speaker-independent continuous-speech recognition on large-sized vocabularies (order of 1000-10,000 words) in the 1980s (Chow et al., 1987; Lee and Rabiner, 1989; Weintraub et al., 1989). This period was also characterized by a shift in methodology from the more intuitive knowledge-based approach toward a more rigorous statistical model framework (Ferguson, 1980; Rabiner, 1989). These methods were actually developed in the 1970s (Baker, 1975; Jelinek, 1976), but their widespread utilization took root in the mid 1980s.

- In the 1990s: Multiple-speaker, large-vocabulary (order of 10,000-20,000 words) continuous-speech recognition under adverse conditions.

  The research of the 1990s was marked by the deployment of real speech-enabled applications for specific/limited tasks, such as user-system dialog in an air travel information application and the extension of ASR systems to very large vocabularies (order of 10,000 and above) for dictation purposes (Makhoul and Schwartz, 1994). Another major trend of this decade was the use of ASR systems within public telephone networks (Wilpon and Rabiner, 1994). As a result, these emerged an increasing interest in speech processing under noisy and adverse conditions, as well as for spontaneous speech recognition.

- In recent years: Multilingual- and multiaccent-speaker, large-vocabulary (above 20,000 words) continuous-speech (conversational/dialog) recognition under adverse conditions.

  Finally, in the last few years, research activities have continued to improve the performance of ASR systems used to recognize very-large-vocabulary continuous speech, dealing with multilingual and multiaccent speakers

(Huang et al., 2004; Bouselmi et al., 2006; Loof et al., 2006), as well as adverse conditions (Matsuda et al., 2006; Morgan and Gold, 1999).

To summarize, the problem of ASR has been approached progressively, from a simple machine that responds to a small set of sounds to a more sophisticated system that responds to real spoken language. Moreover, many methods and algorithms have been developed for ASR. Broadly speaking, those methods can be classified into two main approaches: "knowledge-based" and "corpus-based." The key technologies of both approaches are overviewed in the following section.

1.2 Approaches to Speech Recognition

1.2.1 Knowledge-based Approaches

The idea behind the knowledge-based approaches involves the direct and explicit incorporation of experts' speech knowledge in a rule-based recognition system, as outlined in Figure 1.2.

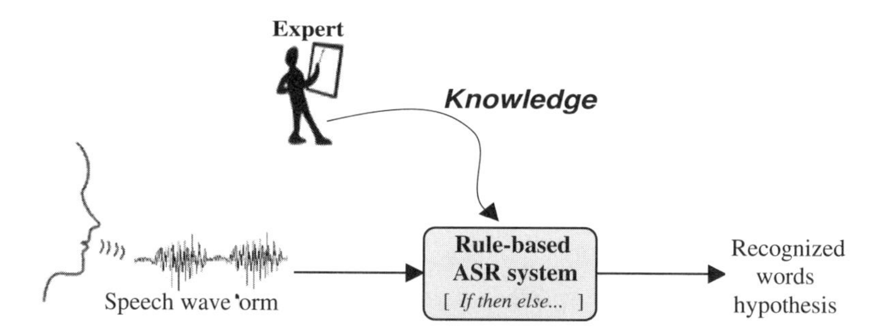

Fig. 1.2. Knowledge-based ASR system.

This knowledge, which is mostly related to the acoustic-phonetic properties of speech signals, is usually derived from a careful analysis of spectrograms (Zue, 1985; Klatt, 1977). A spectrogram is basically presented as a three-dimensional time-frequency-intensity representation of an acoustic signal, which provides a visual display of its relevant temporal and spectral characteristics. Figure 1.3 shows an example of a speech spectrogram reading, which corresponds to the word sequence of "Good night". The dark areas in the spectrogram show high intensity, and the horizontal dark bands show the formant peaks (F1, F2, etc) or the vocal tract resonances. Thus, by visually analyzing the spectrogram, human experts may determine phonetic identities and their realization in a language. This involves multiple knowledge sources, including articulatory movements, acoustic phonetics, phonotactics, and linguistics (Zue and Lamel, 1986).

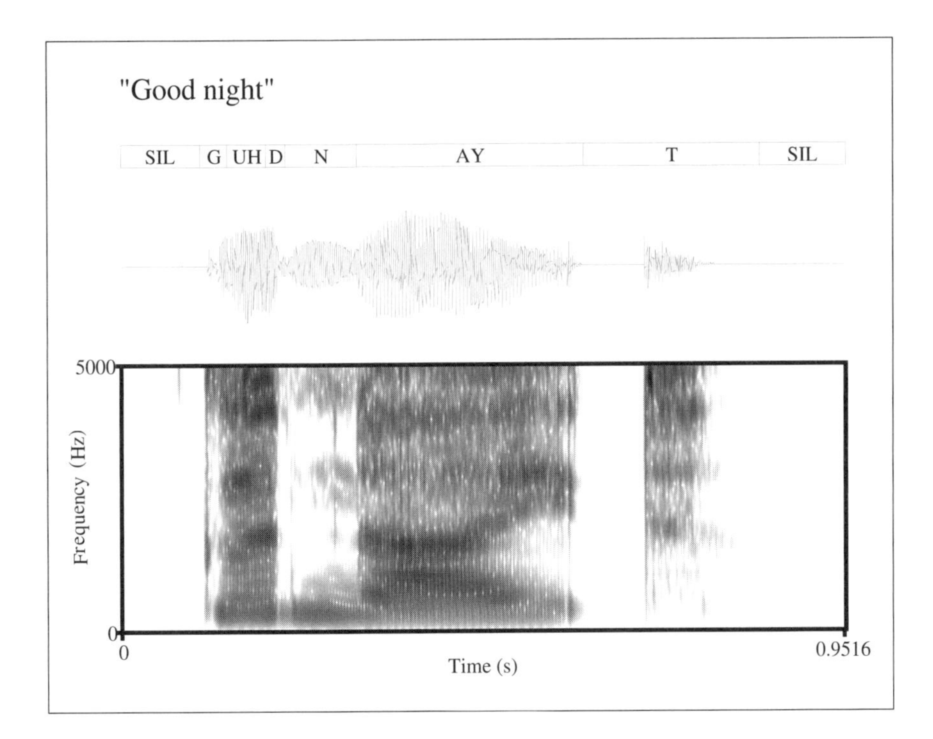

Fig. 1.3. Speech spectrogram reading, which corresponds to the word sequence of "Good night".

While such an approach appears manageable for small vocabularies and isolated words, the task becomes intractable when confronted with the huge number of acoustic-phonetic, lexical, and syntactic characteristics involved in large-vocabulary continuous-speech applications (Waibel and Lee, 1990). As a result, the required computation processes are expensive, and recognition performance has been relatively poor.

In general, knowledge-based approaches face the following major obstacles:

1. Since these approaches highly depend on the ability of human experts to interpret spectrograms, there is an inevitable loss of generality.

2. The knowledge has to be manually entered, so the rules are usually limited in number and scope.

3. As the number of rules increases, inconsistencies may occur among them.

4. Covering a wide range of domains appears to be difficult. Currently, it is estimated that an effort of around 15 years is required to obtain a sufficiently

large amount of knowledge for speaker-independent large-vocabulary continuous speech recognition (Mariani, 1991). Other work (Stern et al., 1986) concluded that, generally, this type of approach only studies a specific set of phonemes for a specific speaker.

1.2.2 Corpus-based Approaches

In contrast, the corpus-based, or statistical, approaches are usually based on modeling the speech signal using well-defined statistical algorithms as outlined in Figure 1.4. This is similar to the process in Figure 1.2, but instead of manually entering the experts' knowledge, the system can automatically learn and extract knowledge from a data corpus.

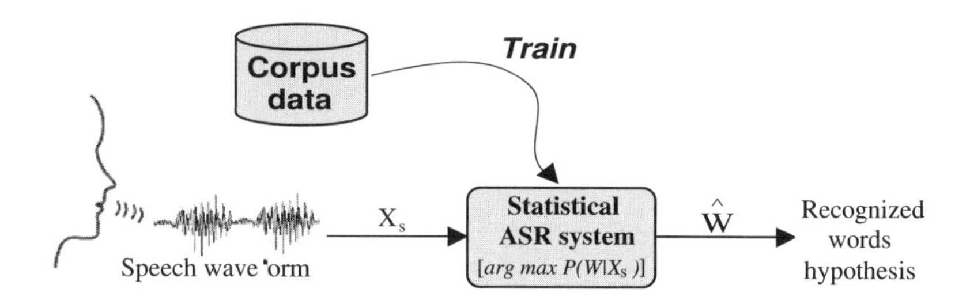

Fig. 1.4. Corpus-based statistical ASR system.

The statistical approaches may also be generally viewed as a way of solving a pattern recognition problem, where the pattern of speech signal X_s is mapped into a set of strings of known words $\hat{W}$. However, since the stochastic variables in the pattern of the speech signal X_s are affected by many factors, pattern recognition is in general more complex than simply deciding on whether an input vector belongs to a known word, and thus the system needs to know the stochastic rules - encoded in the acoustic parameters of the system - that perform the mapping. In practical applications, however, these stochastic rules are never explicitly known. Therefore, the common core of the statistical classification perspective (Schuermann, 1996) is "learning by examples" from a collection of data.

The statistical approaches to continuous-speech recognition involve spectral analysis, acoustic modeling, pronunciation matching, language modeling, and search. Here, the statistical formulation is calculated for the given acoustic observation sequence, $X_s = x_1, x_2, ..., x_T$; from this, the most probable word sequence, $P(W|X_s)$, that a human might utter, $\hat{W} = w_1, w_2, ..., w_n$, is chosen from all word sequences.

This modeling approach has achieved encouraging results, and it outperforms the knowledge-based approach. Today's state-of-the-art ASR systems achieve very good performance in controlled conditions (Pallett, 2003; Juang and Rabiner, 2005; Tolba and O'Shaughnessy, 2001). The following section describes the performance of statistical ASR systems in more detail.

1.3 State-of-the-art ASR Performance

Despite the rapid progress made in statistical speech recognition, many challenges still need to be overcome before ASR systems can reach their full potential in widespread everyday use. The desired goal of a machine that can understand the spontaneous speech of different speakers using various languages in different environments is still far from reality (Tolba and O'Shaughnessy, 2001).

Figure 1.5 depicts the reported results of the best systems on the 2003 NIST's benchmark ASR test history, including read speech and conversational speech (Pallett et al., 1999; Pallett, 2003). The figure shows that error rates on the spontaneous speech portion of the test set were nearly double those of the portion conducted under planned, studio-recorded conditions. Furthermore, in the presence of unexpected acoustic variability, ASR systems often perform much worse than human listeners (Lippmann, 1997; Weintraub et al., 1996).

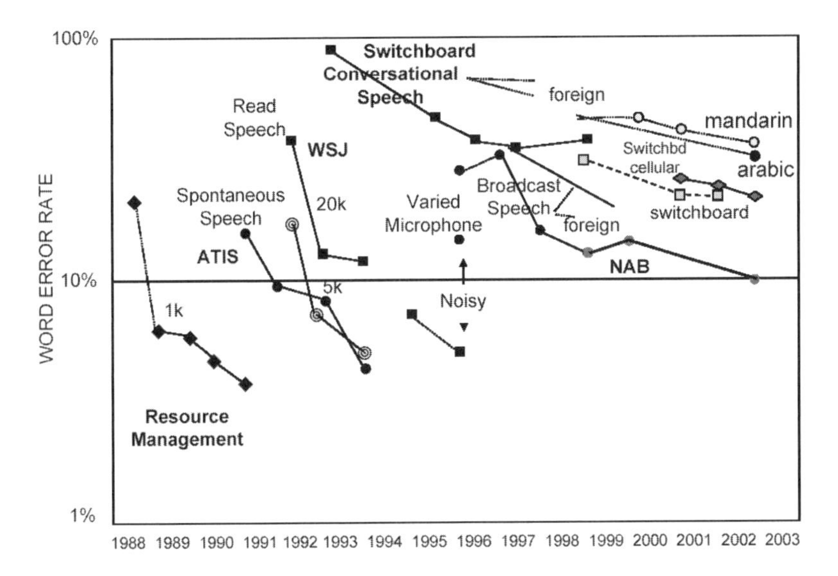

Fig. 1.5. 2003 NIST's benchmark ASR test history (After Pallett, 2003, ©2003 IEEE).

Recently, the TC-STAR (Technology and Corpora for Speech to Speech Translation) ASR evaluation campaign assessed speech recognition performance for three languages (English, Spanish, and Mandarin). It used three evaluation tasks (Lamel et al., 2006; Docio-Fernandez et al., 2006; Loof et al., 2006; Kiss et al., 2006; Ramabdharan et al., 2006):

- European Parliament Plenary Sessions (EPPS). This consisted of audio recordings of the EPPS for English and Spanish, obtained from the original channel of the parliamentary debates.

- CORTES Spanish Parliament Sessions. Since there are few Spanish speeches in EPPS, the audio recordings of the Spanish Parliament (Congreso de Los Diputados) were included.

- Broadcast news. This consisted of audio recordings of broadcast news from Mandarin "Voice of America" (VOA) radio stations.

The reported results for the best systems are outlined in Figure 1.6 (Choukri, 2007). From the performance analyzes of each speaker, it was found that one factor causing performance degradation was the effect of accents and speaking style (Mostefa et al., 2006).

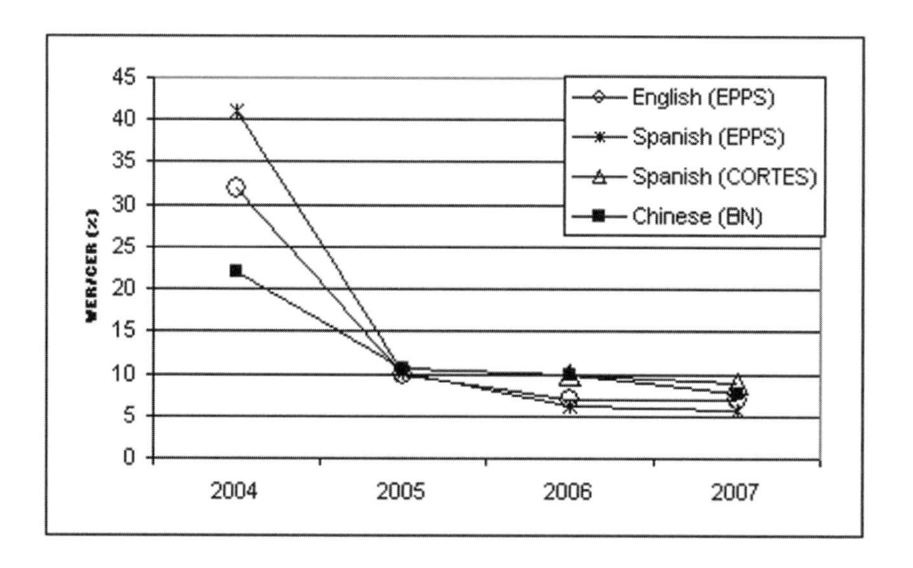

Fig. 1.6. TC-STAR ASR evaluation campaign (After Choukri, 2007, ©TC-STAR).

In summary, the performance of statistical ASR systems drops as the task constraints are relaxed, such as, going from isolated speech to continuous speech, single speaker to multiple speakers, single accent to multi-accent and multilingual, small vocabulary to large vocabulary, and clean conditions to noisy conditions (Doss, 2005). One main reason for such reduced performance is that the statistical ASR systems highly depend on training data, and thus they are not robust to interference that does not exist in the training data. To improve performance, we usually rely on collecting more data to train more detailed models (Li et al., 2005). However, such resources are rarely available, since the presence of variabilities in speech arise from many different factors, and thus a huge amount of training data is required to cover all variabilities. In other words, it is not enough in handling these variabilities by relying solely on statistical models. The systems need more additional knowledge on speech, which could help to handle these sources of variability. Otherwise, only a limited level of success could be achieved. Figure 1.7 illustrates an abstract "S" curve of the progress in current ASR technology (inspired by the ideas from Lee (2004)). Combining advantages of both knowledge-based and statistical approach may be a good candidate for the next generation of ASR technology, and may capable to go beyond the current limitations.

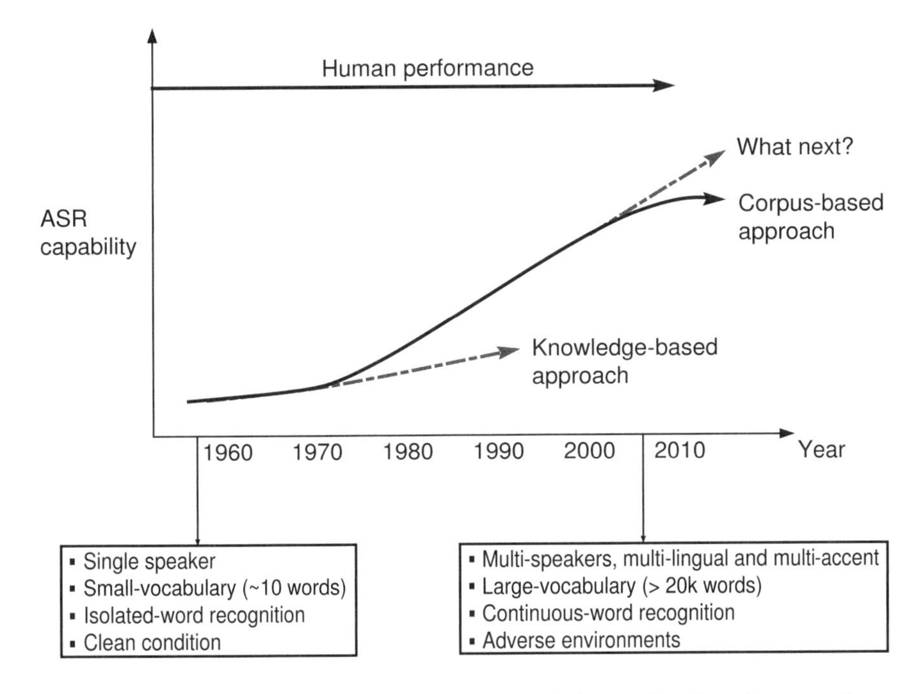

Fig. 1.7. "S" curve of ASR technology progress and the predicted performance from combining deep knowledge with a statistical approach.

Many researchers are aware of the need to incorporate knowledge into current ASR systems, so explicit efforts have been made to integrate knowledge-based and statistical approaches. We explore these existing methods more details in the following section. We first describe different sources of variability that might be present in the speech signal. Then we investigate existing ways of incorporating knowledge sources that have been pursued by other researchers.

1.4 Studies on Incorporating Knowledge Sources

1.4.1 Sources of Variability in Speech

Numerous sources of variability are contained in a speech signal (Holmes and Huckvale, 1994; Huang et al., 2001). The major classes of the variabilities inherent in acoustic speech signals are:

1. Contextual variability

   The acoustic realization of a given phoneme depends on its neighboring context (coarticulation). This is a fundamental part of language sound systems that allows for dynamic transitions between adjacent phoneme segments (both within and across words) and that perhaps even makes speaking easier (Scarborough, 2004). As a result, phonemes can have very different waveforms when produced in the context of other phonemes (Rabiner and Juang, 1993).

   These effects can be produced not only by the first preceding/succeeding contexts but also from more distant neighboring contexts. Research by Scripture (1902) found that a vowel may influence not only the preceding consonant but also the vowel before that consonant. Records of /eli/ and /ela/ or /ebi/ and /eba/ showed that the articulatory setting for /e/ was different according to the second vowel in the sequence: the tongue rose higher and nearer to the /i/ in /eli/ and /ebi/ than in tokens in which the last sound constituted an /a/ (Kuehner and Nolan, 1999). Other studies by Heid and Hawkins (2000), as well as West (2000), found that English consonants such as /l/ and /r/ exert long-distance coarticulation effects across syllables, or "resonance."

2. Speaker variability

   Speaker variability can be further categorized as within-speaker or across-speakers types:

- Intra-speaker (stylistic) variability

  The main sources of variability with a single, individual speaker are commonly called speaking styles. A speaker may modify his/her voice quality, speaking rate, fundamental frequency or even articulatory patterns depending on the physiological and psychological state of the speaker while still transmitting the same linguistic message (Junqua and Haton, 1996). Lexical stress, which is the relative emphasis that may be given to certain syllables in a word, also varies within an individual speaker. A study by Aull and Zue (1985) shows that knowing the stress pattern of a word can greatly reduce the number of competing word candidates.

- Inter-speaker variability

  Speech produced by an individual speaker differs from that of others because it reflects the physical vocal tract, gender, age, accent, and so on. Several researchers (Klatt and Klatt, 1990; Henton, 1992) have reported voice quality differences among male and female speakers. Average female speakers have higher formant frequencies and breathier quality than male speakers. Female speakers typically tend to articulate more than males do (Beinum, 1980). Studies of the effects of age on speech acoustics have also been presented in (Lee et al., 1999; Iseli et al., 2006). Another study (Ghorshi et al., 2006) compares the effect of three different major English accents, namely British, Australian and American. Results show that the formants of the vowel play an important role in conveying the differences between English accents.

3. Channel and environmental variability

   There is a wide variety of noises in everyday-life environments, such as engine noise from vehicles, street traffic noise, and even radio, television, and air conditioning noise. This background noise may mask sounds or even entire words. Changes in duration and amplitude were also found in echoey environments (Howell et al., 1992). As reviewed by Junqua and Haton (1996), a number of studies quantified the acoustic changes induced by noises (Hansen, 1988; Summers et al., 1988). Another source of noises is characterized by channel transmission (e.g., over telephone lines). Telephone speech is more difficult to recognize than high-quality clean speech due to bandwidth limitation, handset and connection quality variations and increased background noise (Junqua and Haton, 1996). Some studies evaluated the problems of ASR specific to telephone speech (Wilpon, 1989; Chigier and Spitz, 1990).

1.4.2 Existing Ways of Incorporating Knowledge Sources

Many techniques have been proposed to incorporate knowledge sources in order to handle the variabilities of speech. These are described as follows.

a. Incorporating contextual information

To deal with contextual variabilities, there is obvious need for LVCSR systems that can accurately capture coarticulation effects. The wider the acoustic unit models, the better the capturing of the coarticulation effects (Pfau et al., 1997). Word unit models are impractical for LVCSR systems due to the large amount of training data needed, the large decoding search space, and the inefficiency of expanding the vocabulary system. Syllable-based (Shafran and Ostendorf, 2000; Ganapathiraju et al., 2001) and multiphone (Messina and Jouvet, 2004) units are smaller than words, both in number and duration, although there are still too many of them and, like words, they lack generality (O'Neill et al., 1998). For example, in the large SWITCHBOARD (SWB) corpus developed by Godfrey et al. (1992), there are about 9,000 syllables appearing in the training database, but over 8,000 of these have fewer than 100 training tokens (Ganapathiraju et al., 2001).

The phonetic units are thus a natural choice since there are only a few of them and their frequency of appearance in the training data is much higher. A standard solution to the coarticulation problem is to extend the phonetic units to include context (Smith et al., 2001). Most of the current LVCSR systems use the context-dependent triphone as the fundamental acoustic unit. Context-dependent triphone units have the same structure as context-independent phonetic (monophone) units, but are trained on data with immediately preceding and following phonetic context information (O'Neill et al., 1998).

Although such triphones have proved to be an efficient choice, it is believed that they are insufficient for capturing all of the coarticulation effects. Some research works reported by Finke and Rogina (1997) and by Bahl et al. (1991) have attempted to improve acoustic models by incorporating a wider-than-triphone context, such as a tetraphone, quinphone/pentaphone, or still larger system. The IBM, Philips/RWTH, and AT&T LVCSR systems have also been quite successful in using pentaphone models (Neti et al., 2000; Beyerlein et al., 1999; Ljolje et al., 2000). However, to properly train the model parameters and to use them in cross-word decoding, large amounts of training data and memory space are usually required. For large-scale systems, a simple procedure to avoid decoding complexity is to apply wide-context models in the rescoring pass. In this case, the decoding will use knowledge sources of progressively increasing complexity to decrease the size of the search space (Hori et al.,

2003). Another possibility is to use only intra-word wide-context units (Beyerlein et al., 1999).

Mohri et al. (2002) and Riley et al. (1997) proposed to compile wide-context-dependent models into a network of Weighted Finite State Transducers (WFST), in order to completely decouple the decoding process from the wide context. However, when higher-order models are used, difficulties lie in the compilation itself. The work by Schuster and Hori (2005) was thus conducted as an attempt to simplify the compilation method.

Furthermore, the incorporation of higher-level linguistic information related to syllable structure and word position, using decision-tree-based acoustic modeling, has also been proposed by Ostendorf (2000), Fosler-Lussier et al. (1999), Reichl and W.Chou (1999) and also Shafran and Ostendorf (2000). Word position information appears to be useful, but information on syllable position leads to small gains.

b. Incorporating speaker information

Several methods have been proposed to incorporate auxiliary information, including pitch frequency (Doss et al., 2003; Doss, 2005), lexical stress (Wang and Seneff, 2001), and contour of the lips (Dupont and Luettin, 2000). Another method proposed by Li et al. (2005) is to classify manners and places of articulation by incorporating sources of acoustic-phonetic knowledge using neural networks for rescoring purposes. The work by Siniscalchi et al. (2006) attempts to build embedded knowledge-based speech detectors for real-time execution.

The most common solution for dealing with inter-speaker variability is to incorporate gender information by building gender-dependent systems (Vergin et al., 1996). These are usually created by splitting the training data into the two genders and building a separate acoustic model for each gender. Another study (Neti and Roukos, 1997) attempted to build phone-specific gender-dependent acoustic models where the gender information was included in addition to phone context questions in the context decision tree.

Some studies have attempted to handle accented speech by modeling phonetic changes where the phoneme set is extended to include accent-specific units (Liu and Pascale, 2006). As a consequence, however, the extended phone set may introduce more lexical confusion in the decoder. Another common technique is to apply adaptation techniques (Huang et al., 2000; Wang et al., 2003; Leggetter and Woodland, 1995) to modify acoustic model parameters to fit the characteristics of a particular accents. However, the parameters of acoustic models undergo an irreversible change, and the models lose their ability to cover other accents. Other studies then prefer to build new recognizers for specific accent regions, either to be used independently as accent-dependent models (Beattie et al., 1995) or in conjunction with an accent-switch (Kumpf and King, 1997).

c. Incorporating channel or environmental information

A review by Gong (1995) described the various techniques that have been used to improve robustness against noise. Basically, we may remove noise from the signal (speech enhancement) or transform speech models to accommodate noise. These techniques are called feature-based methods and model-based methods, respectively. Since this book focuses on incorporating knowledge sources into the model, these feature-based techniques are not discussed here.

Acoustic modeling techniques for improving noise robustness may be done by multi-condition training, where a single acoustic model is estimated with a large database that contains several environments including different noise types and noise levels. However, Matsuda et al. (2006) showed that the variety of speaking environments for which a single model can have high performance is limited. One solution is to employ multiple acoustic models, one model for each different condition. Their experiments revealed that it could achieved higher recognition performance than the single model.

Other methods have been use to adapt acoustic models to particular environmental conditions. These techniques include PMC (Gales and Young, 1992, 1995), NOVO (Martin et al., 1993), and MLLR (Leggetter and Woodland, 1995). However, these methods tend to require a considerable amount of noise-sample training data and too much computation to allow real-time monitoring of instantaneous changes in the noise spectrum. A study by Sagayama et al. (1997) proposed a fast adaptation method using a Jacobian matrix.

The above methods and techniques were mainly proposed to handle particular sources of variability (i.e., robust to coarticulation effects only or robust to the noise environment only). Although they have proven to be useful, difficulties arise when several types of knowledge sources need to be incorporated.

Recently, the use of probabilistic graphical methods such as Bayesian networks (BNs) in ASR, has gained attention (Bilmes, 1999; Zweig and Russell, 1998). It appears that dynamic BNs (DBN), which can be regarded as a generalization of hidden Markov models (HMMs), can be used to easily incorporate various additional types of knowledge, such as auxiliary Information, sub-band correlation, and speaking rates (Stephenson et al., 2001; Daoudi et al., 2000; Shinozaki and Furui, 2000). This alternative modeling approach seems to be an ideal candidate for incorporating various types of knowledge sources. However, when many knowledge sources are incorporated, the computational complexity and memory requirements of model inference tends to increase exponentially in the number of nodes. In such cases, the models often become impracticable. Consequently, the lack of efficient techniques has limited the application of BNs to large-vocabulary continuous-speech recognition (LVCSR) in the past. The work reported in this book attempts to resolve these issues, the most critical of which are set out in the following section.

1.4.3 Major Challenges to Overcome

Generally speaking, incorporating various additional knowledge sources leads to improvements in the recognition rate, but there have often been cases where developing such models and achieving optimal performance have not been feasible. This situation might be due to the following reasons. First, the models are robust only to a particular source of variability, having little or no ability to handle other factors. Second, the models may be able to handle various sources of variability, but thus become too complicated. When there are insufficient resources for proper training of model parameters (i.e., the amount of training data and memory space available), input-space resolution may be lost due to non-robust estimates and the increasing number of unseen patterns. Moreover, decoding with large models may also become cumbersome and sometimes even impossible.

Therefore, developing an efficient modeling method that can maintain a delicate balance, between obtaining various kind of deep knowledge and the feasibility of the training and recognition effort, is one of the most important problems to overcome for a realistic application of ASR systems. Here, we outline three major challenges that have to be faced when developing such systems:

- How can we incorporate various knowledge sources from different domains in an efficient and unified way?
- At which level of an ASR system can we incorporate these additional knowledge sources?
- How can we solve the model complexity issues arising from the incorporation of a significant number of additional knowledge sources? These issues include the availability of training data, as well as the training and recognition effort.

1.5 Book Outline

The book addressed a problem of developing a novel ASR technology that can exploit wide-ranging knowledge of speech variability while keeping the training and recognition effort to a feasible cost, leading to the delivery of solutions for the major challenges described above. Naturally, this work aims to achieve this goal while improving the robustness of speech recognition in term of accuracy.

Therefore, in this book, a new efficient general framework to incorporate additional knowledge sources in state-of-the-art statistical ASR systems (Figure 1.8) is provided, investigated and evaluated. Since it is based on a graphical model, the framework is called GFIKS (graphical framework to incorporate knowledge sources).

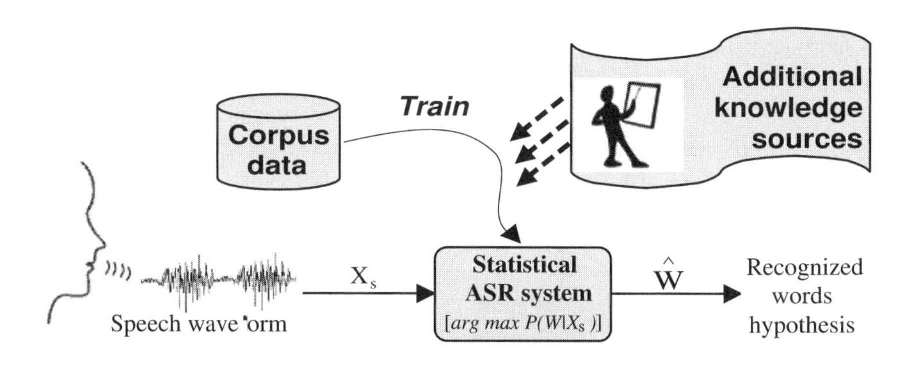

Fig. 1.8. Incorporating knowledge into a corpus-based statistical ASR system.

This book is organized as described below.

In Chapter 2, we describe the state-of-the-art statistical ASR technology based on HMMs. After an overview of pattern recognition, we briefly review the basic theory of HMMs, including the fundamental theory of Markov chains, the general form of HMMs, and three practical problems in using HMMs for ASR systems. Finally, we describe the pattern recognition task for HMM-based ASR systems, including feature extraction, HMM-based acoustic modeling, language modeling, pronunciation dictionary and the search algorithm. Descriptions of phone-unit-dependent models, speech observation density, and various approaches to parameter tying are also provided.

In Chapter 3, we introduce the design of our proposed GFIKS framework. Since this framework is based on a graphical model, we first provide a review of Bayesian statistical theory and graphical model representation, including the Bayesian network and junction tree decomposition. Then, we introduce a procedure for incorporating knowledge sources, including defining the causal relationships between information sources, performing inference on a BN, and also finding solutions when direct inference is intractable. Finally, we discuss the general issues and potential of incorporating knowledge sources in an HMM acoustic model. This includes the type of knowledge sources to be incorporated and the level of HMMs where these sources need to be incorporated.

In Chapter 4, we show how to incorporate additional knowledge sources in a statistical speech recognition system by utilizing GFIKS. First, we incorporate additional sources of knowledge at the HMM state level. Then, we incorporate additional sources of knowledge at the HMM phonetic unit level. The descriptions at both levels include acoustic model topology, inference, training issues, and parameter reduction. We also present experimental results by incorporating various knowledge sources, including noise, accent, gender and wide-phonetic context information.

In Chapter 5, the conclusions of the book are presented. Future directions are also discussed, focusing on how to incorporate the developed approaches in spoken language dialog systems.

2

Statistical Speech Recognition

This chapter describes the state-of-the-art technology for statistical ASR based on the pattern recognition paradigm. The most widely used core technology is the hidden Markov model (HMM). This is basically a Markov chain that characterizes a speech signal in a mathematically tractable way.

Section 2.1 provides an overview of pattern recognition. In Section 2.2, we review the theory of Markov chains and the general form of an HMM, including three practical problems in using HMMs. In Section 2.3, we describe in detail the pattern recognition task for HMM-based ASR systems, starting from feature extraction, which processes the speech signal into a set of feature patterns, up through the search algorithm, which maps those features into the most probable strings of words. We also explain language modeling, the pronunciation dictionary, and acoustic modeling, including phone-unit-dependent models, speech observation density, and various approaches to parameter tying.

2.1 Pattern Recognition Overview

A variety of technological approaches to speech recognition have been developed within the context of a long history of pattern recognition studies. Although specific methods continue to evolve, pattern recognition remains a useful perspective for describing many problems and their solutions.

As used here, the word *pattern* means something that exhibits certain regularities, something that can serve as a model or an object, or something that represents the concept of what is observed. The word *recognition* refers to the task of understanding in a meaningful way the kind of object we have observed (Schuermann, 1996).

Pattern recognition may also be described in terms of successfully classifying a set of measurements into categories. The fundamental approach is to consider the pattern as having two different linked worlds that belong to each

S. Sakti et al., *Incorporating Knowledge Sources into Statistical Speech Recognition*,
Lecture Notes in Electrical Engineering 42, DOI: 10.1007/978-0-387-85830-2_2,

other like the two sides of a coin: A world of physical observations and a world of concepts and ideas.

Using the fundamental notions of pattern classification (Schuermann, 1996), there exist :

- a world of physical observations x,

$$x = [x_1, x_2, ..., x_n, ..., x_N], \tag{2.1}$$

  where N is the number of measurements taken, and
- a world of possible target categories or classes y

$$y = [y_1, y_2, ..., y_k, ..., y_K], \tag{2.2}$$

  where the K classes are mutually exclusive and complete.

From this viewpoint, we may consider the pattern a pair of variables comprising an observation and a meaning,

$$\text{Pattern} = [x, y], \tag{2.3}$$

and thus the task of pattern classification may be considered as the mapping of $x \rightarrow y$.

In a mathematical sense, the two worlds of physical observations, on the one hand, and concepts, names, and meanings, on the other, correspond to two spaces. Designing a pattern recognition system amounts to establishing a mapping of $x \rightarrow y$ from a measurement space X into a decision space Y, containing K discrete points, each representing one of the K classes. Figure 2.1 shows an example of mapping multi-dimensional measurement space X to three-class target vectors in decision space Y.

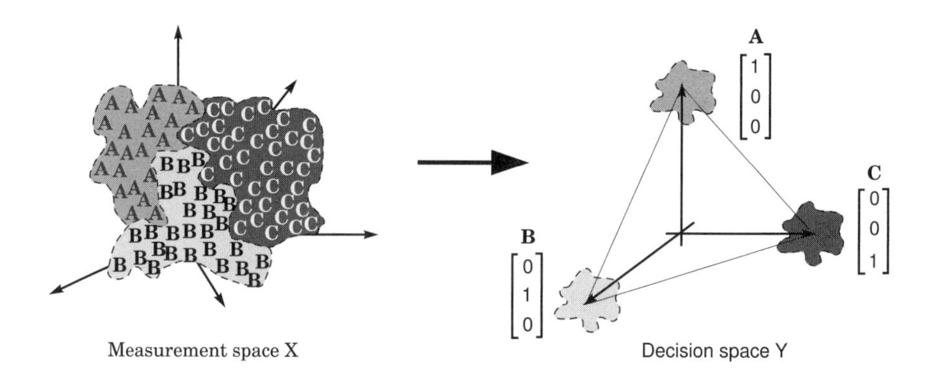

Fig. 2.1. Pattern recognition: Establishing mapping from multi-dimensional measurement space X to three-class target decision space Y.

The mapping $x \rightarrow y$ should, in as many cases as possible, match the corresponding vector in Y, representing one of the K classes. However, the difficulty of the classification task, expressed in terms of the minimum achievable error rate, depends on whether these distributions are locally concentrated or spread out over a wide region of X, and specifically on the degree of their mutual penetration. Since pattern recognition is in general more complex than simply deciding on whether an input vector belongs to one of the K classes, the statistical classification perspective is useful (Schuermann, 1996). In this case, we have an underlying probability model, which tells us the optimum classifier with a given probability of being in each class rather than simply a classification.

Using the fundamental notation of the statistical framework, we denote the probability of a measurement vectors x_n belonging to class y_k as:

$$p(y_k|x_n) \quad \text{where} \quad 0 \leq p(y_k|x_n) \leq 1 \quad \text{and} \quad \sum_{k=1}^{K} p(y_k|x_n) = 1. \tag{2.4}$$

Since this probability can only be estimated after the data has been seen, it is generally referred to as the posterior or *a posteriori* probability of class y_k (Bourlard and Morgan, 1994). Consequently, we can find the optimum decision that assigns x_n to class y_k if

$$p(y_k|x_n) > p(y_j|x_n), \quad \text{where} \quad \forall j = 1, 2, \ldots, K, \quad \text{and} \quad j \neq k. \tag{2.5}$$

This optimum strategy is often called "Bayes" decision rule, which assigns the class y_k that yields the highest posterior probability, given the measurement vector x_n (Fukunaga, 1990).

The pattern itself is represented by a stochastic variable, and there are a variety of different patterns, each yielding a certain probability of being observed (Schuermann, 1996). A pattern consists of a discrete number of self-identical objects provided with a number of fixed or very slowly changing attributes. Some of these represent the peculiarity of a specific object; the remainder, which may be called "features," determine the class the object belongs to. Therefore, this mapping must take into account knowledge on the variabilities of patterns, which characterize the vector x for each of the classes y it may be assigned to. In other words, it is necessary to determine the correct meaning y of those patterns that are similar to the given examples of a feature x.

From this point of view, it makes no difference what type of observations are considered and to what meaning they may be linked. The same approach applies for recognizing written text, spoken language, camera images, or any other type of multidimensional signal interpretation. For our purposes, we adopt the pattern recognition approach for ASR, we substitute "x" with a "speech signal of spoken utterance" and "y" with "strings of words." Then, the pattern recognition approach is used to map a speech signal into a set of meaningful strings of words (see Figure 2.2).

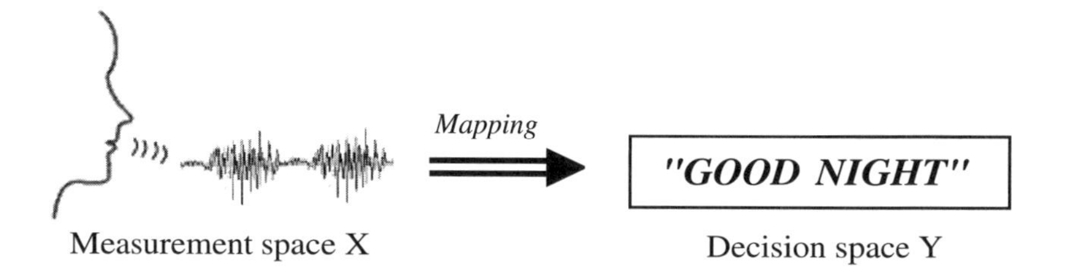

Fig. 2.2. Pattern recognition approach for ASR: Establish mapping from measurement space X of speech signal to target decision space Y of word strings.

Before further discussing the matter of pattern recognition approach for ASR in more detail, we first review the HMM theory in the next section.

2.2 Theory of Hidden Markov Models

2.2.1 Markov Chain

A Markov chain is a finite-state system with stochastic transitions, yielding associated probabilities (Gold and Morgan, 1999). The output of each Markov state corresponds to a deterministic event.

Let $W = \{sunny, cloudy, rainy\}$ be the deterministic event of the weather. Then, a simple example of a three-state Markov chain, in which each Markov state corresponds to one weather event, is shown in Figure 2.3.

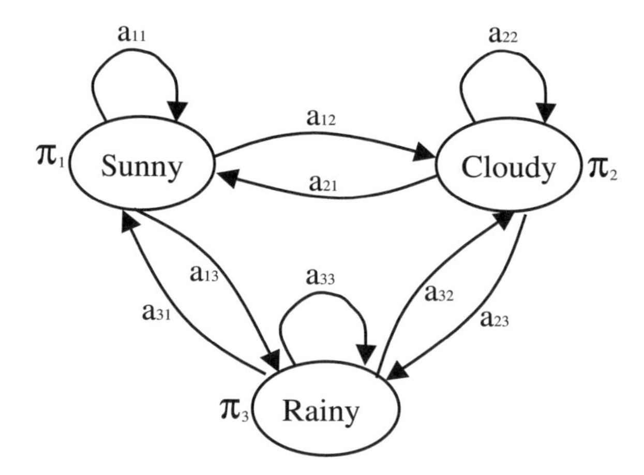

Fig. 2.3. Simple three-state Markov chain for daily weather.

The Markov state at time t is denoted as q_t, with a stochastic state-transition probability matrix:

$$A = \{a_{ij}\} = P(q_t = j|q_{t-1} = i) = \begin{bmatrix} a_{11} & a_{12} & a_{13} \\ a_{21} & a_{22} & a_{23} \\ a_{31} & a_{32} & a_{33} \end{bmatrix}$$

and initial state probability matrix:

$$\pi = \{\pi_i\} = P(q_1 = i) = \begin{bmatrix} \pi_1 \\ \pi_2 \\ \pi_3 \end{bmatrix}.$$

The probability $P(q_1, q_2, ..., q_T)$ of a possible sequence of weather on day 1,2,..., until day T is calculated as:

$$P(q_1, q_2, ..., q_T) = P(q_1) \prod_{t=2}^{T} P(q_t|q_1, q_2, ..., q_{t-1}). \tag{2.6}$$

Using the first-order Markov assumption, the probability of the random variable at a given time depends only on the value at the preceding time, and thus Eq. (2.6) may be simplified as:

$$P(q_1, q_2, ..., q_T) = P(q_1) \prod_{t=2}^{T} P(q_t|q_{t-1}) = \pi_i \prod_{t=2}^{T} a_{ij}. \tag{2.7}$$

More details on the basic theory of Markov chains may be found in (Gold and Morgan, 1999; Meyn and Tweedie., 1993; Booth, 1967).

2.2.2 General form of an HMM

An HMM represents an extension of a Markov Model, where each Markov state corresponds to a non-deterministic event with an associated observation probability and where the generating state sequence becomes unobserved or hidden (Huang et al., 2001).

An HMM may also be viewed as a double stochastic process, since for each point in time the process undergoes a change of state according to a set of state transition probabilities. After each transition, the process produces a symbol of the state according to an observation probability (Rabiner and Juang, 1993).

Formally speaking, the elements of an HMM are:

1. A set of a finite number of states
   $\Omega = 1, 2, ..., N$; where N is the total number of states,
   and a state sequence of time length T is denoted as
   $Q = q_1, q_2, ..., q_T$; where q_t is the state at time t.

2. A set of distinct observation symbols for each state that correspond to the physical output of the system being modeled.
   $O = 1, 2, ..., M$; M = total number of symbols,
   and the observed output sequence of time length T is denoted as
   $X_s = x_1, x_2, ..., x_T$; where x_t = the observed output at time t.
3. The transition probabilities from state i to state j
   $A = \{a_{ij}\}$; where $a_{ij} = P(q_t = j|q_{t-1} = i)$.
4. The observation symbol probability distribution in state j
   $B = \{b_j(x_t)\}$; where $b_j(x_t) = P(x_t|q_t = j)$.
5. The initial state distribution
   $\pi = \{\pi_i\}$; where $\pi_i = P(q_1 = i)$.

It can be seen that a complete definition of an HMM requires specification of two model parameters (N and M), specification of observation symbols, and specification of three probability measures A, B, and π. For convenience, we use the compact notation

$$\lambda = (A, B, \pi) \tag{2.8}$$

to indicate the complete parameter set of the model.

An example HMM for the weather problem is shown in Figure 2.4. It is similar to the one presented in Figure 2.3, but there is no longer a one-to-one correspondence between the Markov state and the output symbol.

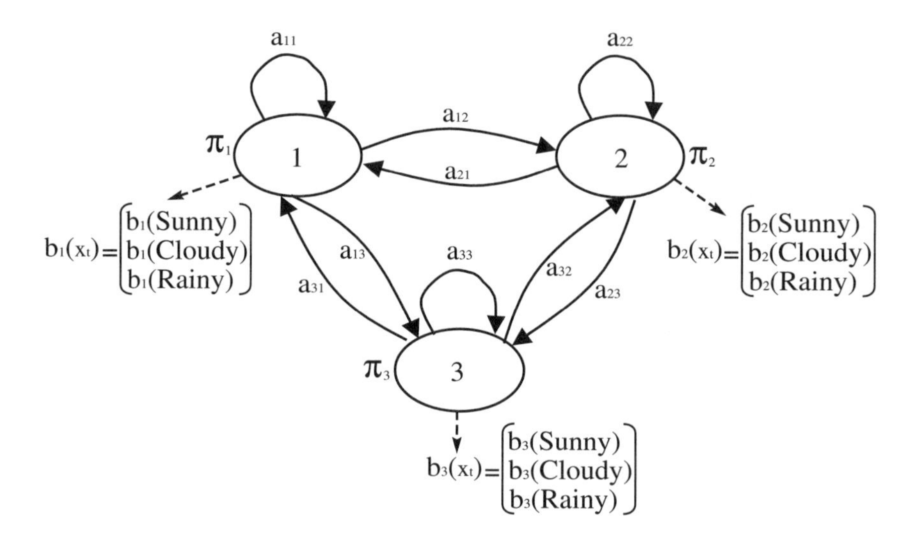

Fig. 2.4. HMM of the daily weather, where there is no deterministic meaning on any state.

Each Markov state can now generate all three output symbols: {*sunny*, *cloudy*, *rainy*}, according to its observation probability function

$$B = \{b_j(x_t)\} = P(x_t|q_t = j) = \begin{bmatrix} b_j(sunny) \\ b_j(cloudy) \\ b_j(rainy) \end{bmatrix}.$$

If the index of the Markov states increases or remains unchanged as time increases, leading to a move from left to right on the chain shown in Figure 2.5, the model is called a left-to-right or Bakis HMM model.

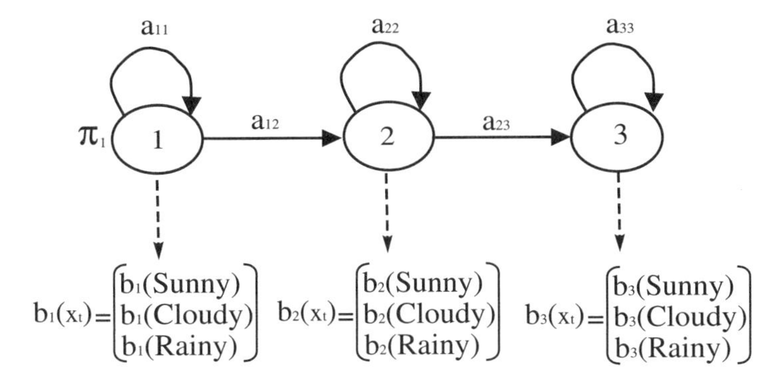

Fig. 2.5. Left-to-right HMM of the daily weather.

2.2.3 Principle Cases of HMM

There are three principal cases that must be dealt with when using an HMM (Huang et al., 2001; Rabiner and Juang, 1993). Each of these are reviewed in the following.

1. **Evaluation issue: How to estimate the HMM total likelihood** $P(X_s|\lambda)$

   Given an HMM λ, the estimation of the probability (likelihood), $P(X_s|\lambda)$, of the observation sequence X_s can be solved as follows:

   - **Direct calculation for full length time** T

     The most straightforward way consists of enumerating all possible state sequences $Q = (q_1, q_2, ..., q_T)$, that generate the observation sequence $X_s = (x_1, x_2, ..., x_T)$, and then summing all of the probabilities:

$$P(X_s|\lambda) = \sum_Q P(X|Q,\lambda)P(Q|\lambda)$$
$$= \sum_Q P(X|Q)P(Q|\lambda), \quad (2.9)$$

where $P(X|Q)$ is the probability of observation variable X that can be derived from all possible $B = \{b_{q_t}(x_t)\}$, and $P(Q|\lambda)$ is the probability of state Q that can be derived from all possible $A = \{a_{q_{t-1}q_t}\}$. Thus, we get

$$P(X_s|\lambda) = \sum_Q [b_{q_1}(x_1)b_{q_2}(x_2)...b_{q_T}(x_T)][\pi_{q_1}a_{q_1q_2}a_{q_2q_3}...a_{q_{T-1}q_T}]$$
$$= \sum_Q \pi_{q_1}b_{q_1}(x_1)a_{q_1q_2}b_{q_2}(x_2)...a_{q_{T-1}q_T}b_{q_T}(x_T)$$
$$= \sum_Q \prod_{t=1}^{T} a_{q_{t-1}q_t}b_{q_t}(x_t), \quad (2.10)$$

where $a_{q_0q_1}$ denotes π_{q_1}.

Figure 2.6 shows the process flow on the trellis diagram of a 3-state HMM with time length T. At each point in time $t = 1, 2, ..., T$, there are N possible states that can be reached. Therefore, this involves an enumeration of $O(N^T)$, and the total number of possibilities increases exponentially with the increasing number of states and observation instances. Consequently, the use of this algorithm seems to be impracticable.

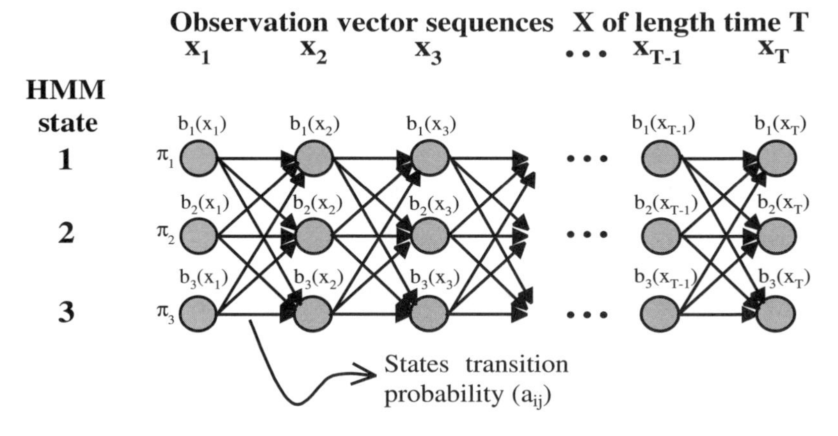

Fig. 2.6. Process flow on trellis diagram of 3-state HMM with time length T.

- **Recursive forward calculation**

A more efficient algorithm, the so-called *forward algorithm*, can be used to calculate $P(X_s|\lambda)$. Its approach is to store intermediate results and to use them for subsequent state-sequence calculations in order to save computational resources (Huang et al., 2001).

Let us define a forward probability $\alpha_t(j)$, which is the probability that the HMM is in state j at time t having generated partial observation x_1^t (namely $x_1, x_2, ..., x_t$):

$$\alpha_t(j) = P(x_1^t, q_t = j|\lambda). \tag{2.11}$$

Then, the complete likelihood $P(X_s|\lambda)$ can be calculated from $t = 1$ with this forward recurrence:

$$\alpha_t(j) = \begin{cases} \pi_j b_j(x_1); & \text{for } t = 1;\ 1 \le j \le N \\ \left[\sum_{i=1}^{N} \alpha_{t-1}(i) a_{ij}\right] b_j(x_t); & \text{for } 2 \le t \le T;\ 1 \le j \le N, \end{cases} \tag{2.12}$$

which terminates in the final column:

$$P(X_s|\lambda) = \sum_{j=1}^{N} \alpha_T(j).$$

Figure 2.7 shows the induction step graphically. Since the probability at each cell $t-1$ has been computed before proceeding to t, the complexity for the forward algorithm is only $O(N^2T)$, rather than exponential.

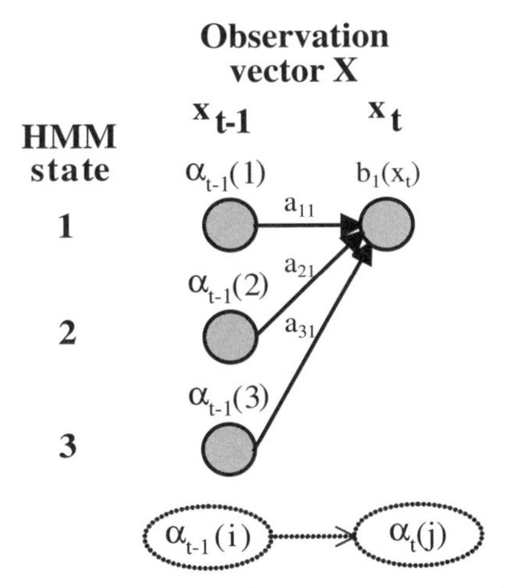

Fig. 2.7. Forward probability function representation (for j=1).

- **Recursive backward calculation**

In a similar manner, we may consider backward calculation to determine $P(X_s|\lambda)$. This is the so called *backward algorithm.* A backward probability $\beta_t(i)$ is the probability that the HMM is in state i at t, having generated a partial observation x_{t+1}^T (from $t+1$ to the end T):

$$\beta_t(i) = P(x_{t+1}^T|q_t = i, \lambda). \tag{2.13}$$

Then, the complete likelihood $P(X_s|\lambda)$ can be calculated from $t = T$ using this backward recurrence:

$$\beta_t(i) = \begin{cases} 1; & \text{for}\, t = T;\ 1 \leq i \leq N \\ \left[\sum_{j=1}^{N} a_{ij} b_j(x_{t+1}) \beta_{t+1}(j)\right]; & \text{for}\, t = T-1, ..., 1;\ 1 \leq i \leq N, \end{cases} \tag{2.14}$$

which terminates in the first column:

$$P(X_s|\lambda) = \sum_{i=1}^{N} \beta_1(i) \pi_i b_i(x_1).$$

Figure 2.8 shows the induction step graphically. Similar to the forward algorithm, the complexity of the backward algorithm is $O(N^2T)$.

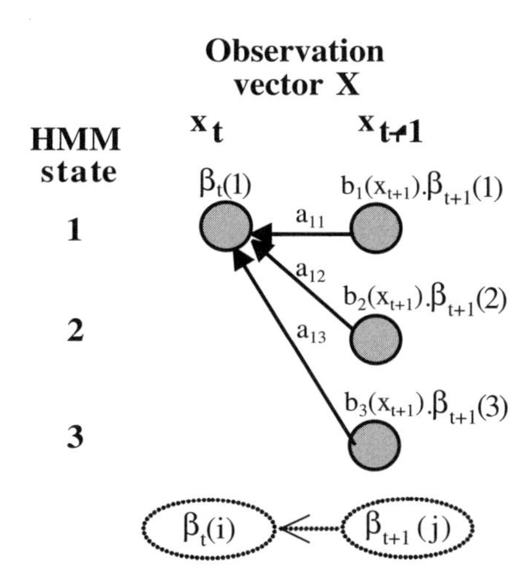

Fig. 2.8. Backward probability function representation (for i=1).

2. **Decoding issue: How to find the optimum HMM state sequence**

A formal technique to determine the single optimum $Q = (q_1, q_2, ..., q_T)$ state sequence exists, and it is the so-called *Viterbi algorithm* (Viterbi, 1967; Fomey, 1973). This technique can also be regarded as a modified forward algorithm, since instead of summing up probabilities from different paths, the Viterbi algorithm chooses and only remembers the best path (Huang et al., 2001). In other words, it simply requires modifying the forward recurrence by replacing all of the summations with the max function.

Let us define the best path probability $\delta_t(j)$, which is the most likely state sequence probability at t, having generated partial observation x_1^t (until t) and ending in state j:

$$\delta_t(j) = P(x_1^t, q_1^{t-1}, q_t = j|\lambda), \tag{2.15}$$

and $\psi_t(j)$ is the array used to keep track of it.

Then, the complete procedure for finding the optimum state sequence can be calculated recursively as follows:

- for the $\delta_t(j)$ probability:

$$\delta_t(j) = \begin{cases} \pi_j b_j(x_1); & \text{for } t = 1;\ 1 \leq j \leq N \\ \max\limits_{1 \leq i \leq N} [\delta_{t-1}(i) a_{ij}]\, b_j(x_t); & \text{for } 2 \leq t \leq T;\ 1 \leq j \leq N \end{cases} \tag{2.16}$$

- for the $\psi_t(j)$ probability:

$$\psi_t(j) = \begin{cases} 0; & \text{for } t = 1;\ 1 \leq j \leq N \\ \arg\max\limits_{1 \leq i \leq N} [\delta_{t-1}(i) a_{ij}]; & \text{for } 2 \leq t \leq T;\ 1 \leq j \leq N, \end{cases} \tag{2.17}$$

which terminates in the final column:

$$\text{The best score} = \max_{1 \leq j \leq N} [\delta_T(j)],$$

$$\hat{q_T} = \arg \max_{1 \leq j \leq N} [\psi_T(j)].$$

By backtracking the best state for each time unit t using:

$$\hat{q_t} = \psi_{t+1}(\hat{q_{t+1}}); \quad t = T-1, T-2, ..., 1$$

we obtain the optimum state sequence:

$$\hat{Q} = (\hat{q_1}, \hat{q_2}, ..., \hat{q_T}).$$

Figure 2.9 shows an example of finding the best path on a trellis diagram. The Viterbi algorithm also yields the complexity of $O(N^2T)$.

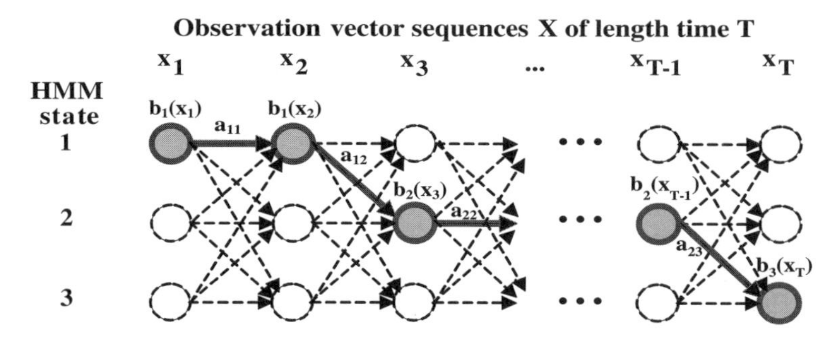

Fig. 2.9. Example of finding the best path on a trellis diagram using the Viterbi algorithm.

3. **Training issue: How to adjust the HMM parameters**

Given a finite observation data sequence X_s, and an underlying state sequence $Q = (q_1, q_2, ..., q_T)$ that is considered to be hidden or unobserved, we try to find the HMM parameters $\lambda = (A, B, \pi)$ that maximize the likelihood of the observed data. The incomplete-data likelihood function is given by $P(X_s|\lambda)$ whereas the complete-data likelihood function is $P(X_s, Q|\lambda)$ (Bilmes, 1998).

A general approach to iterative computation of the maximum-likelihood estimates, when the observations are viewed as incomplete data, is the so-called expectation-maximization (EM) algorithm (Dempster et al., 1977) (also often referred to as the *Baum-Welch algorithm* (Baum, 1972)).

The EM algorithm formally consists of the following two steps:

a) **E-step**: Determine the auxiliary function $G(\lambda, \lambda^m)$, which is the conditional expectation of the complete data likelihood $P(X_s, Q|\lambda)$ with respect to the unknown data Q given the observed data X_s and the current parameter estimates λ^m. That is, we define:

$$G(\lambda, \lambda^m) = E_Q \left[logP(X_s, Q|\lambda) | X_s, \lambda^m \right]. \tag{2.18}$$

b) **M-step**: Calculate a new parameter λ^{m+1} that maximizes $G(\lambda, \lambda^m)$. That is, we find:

$$\lambda^{m+1} = \arg\max_{\lambda} G(\lambda, \lambda^m). \tag{2.19}$$

The log likelihood is often used since it is analytically easier, and the maximum of the log likelihood that also has to be the maximum of the likelihood itself.

The graphical interpretation of the EM algorithm is outlined in Figure 2.10. The goal is to find the global maximum λ of the likelihood $L(\lambda) = P(X_s|\lambda)$. However, since the likelihood function can be arbitrary, the global maximum is difficult to obtain. Instead, by iteratively constructing the corresponding auxiliary function $G(\lambda, \lambda^m)$ and maximizing it, we may climb the surface of $L(\lambda)$ to arrive at local optimum.

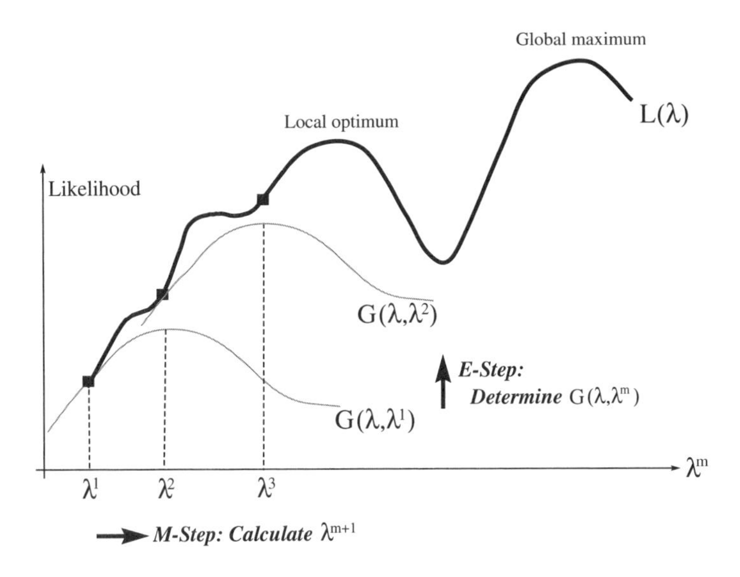

Fig. 2.10. Graphical interpretation of the EM algorithm.

Before describing the calculations in more detail, we define the variable $\xi_t(i,j)$ as the probability of taking the transition from state i at $t-1$ to state j at t. This is the so-called forward-backward probability function (see Figure 2.11):

$$\xi_t(i,j) = P(q_{t-1} = i, q_t = j|X_s, \lambda) = \frac{P(X_s, q_{t-1} = i, q_t = j|\lambda)}{P(X_s|\lambda)}$$
$$= \frac{\alpha_{t-1}(i)a_{ij}b_j(x_t)\beta_t(j)}{\sum_{i=1}^{N}\sum_{j=1}^{N}\alpha_{t-1}(i)a_{ij}b_j(x_t)\beta_t(j)}. \tag{2.20}$$

Then we also define $\gamma_t(i)$ as the probability of being in state i at time t, as

$$\gamma_t(i) = \sum_{j=1}^{N}\xi_t(i,j). \tag{2.21}$$

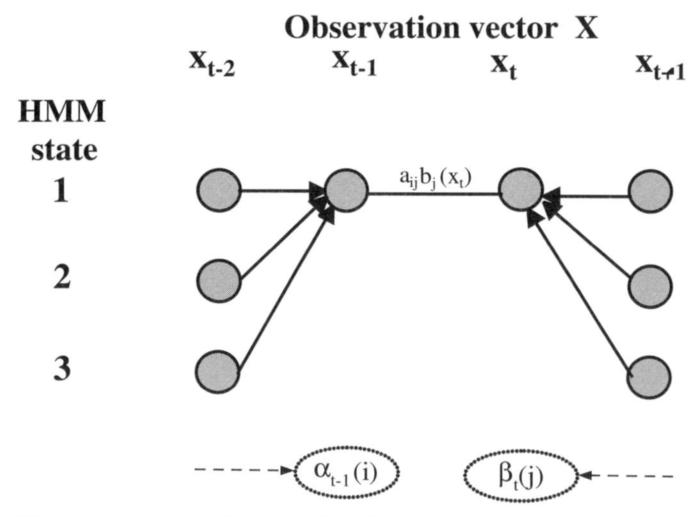

Fig. 2.11. Forward-backward probability function representation.

The details of the HMM parameter calculation are described in the following.

- **E-step: Determine the $G(\lambda, \lambda^m)$ given X_s and λ^m**

  Following Eq. (2.18), the G function is determined as:

$$\begin{aligned} G(\lambda, \lambda^m) &= E_Q\left[logP(X_s, Q|\lambda)|X_s, \lambda^m\right] \\ &= \sum_Q logP(X_s, Q|\lambda)P(Q|X_s, \lambda^m) \\ &= \sum_Q logP(X_s, Q|\lambda)\frac{P(X_s, Q, |\lambda^m)}{P(X_s|\lambda^m)}. \end{aligned} \quad (2.22)$$

  In terms of the HMM parameter as described in Eq. (2.10), $P(X_s, Q|\lambda)$ may be expressed as

$$P(X_s, Q|\lambda) = \prod_{t=1}^{T} a_{q_{t-1}q_t} b_{q_t}(x_t), \quad (2.23)$$

  where $a_{q_0 q_1}$ denotes π_{q_1}, and thus

$$logP(X_s, Q|\lambda) = log\,\pi_{q_1} + \sum_{t=2}^{T} log\,a_{q_{t-1}q_t} + \sum_{t=1}^{T} log\,b_{q_t}(x_t). \quad (2.24)$$

  The G function then becomes:

$$G(\lambda, \lambda^m) = G(\pi_i, \lambda^m) + G(a_{ij}, \lambda^m) + G(b_j, \lambda^m), \tag{2.25}$$

where

$$G(\pi_i|\lambda^m) = \sum_Q [log\, \pi_{q_1}] \frac{P(X_s, Q, |\lambda^m)}{P(X_s|\lambda^m)} = \sum_{i=1}^N log\, \pi_i \frac{P(X_s, q_1 = i|\lambda^m)}{P(X_s|\lambda^m)},$$

$$G(a_{ij}, \lambda^m) = \sum_Q \left[\sum_{t=2}^T log\, a_{q_{t-1}q_t}\right] \frac{P(X_s, Q, |\lambda^m)}{P(X_s|\lambda^m)}$$
$$= \sum_{i=1}^N \sum_{j=1}^N \sum_{t=2}^T log\, a_{ij} \frac{P(X_s, q_{t-1} = i, q_t = j|\lambda^m)}{P(X_s|\lambda^m)},$$

$$G(b_j, \lambda^m) = \sum_Q \left[\sum_{t=1}^T log\, b_{q_t}(x_t)\right] \frac{P(X_s, Q, |\lambda^m)}{P(X_s|\lambda^m)}$$
$$= \sum_{j=1}^N \sum_{t=1}^T log\, b_j(x_t) \frac{P(X_s, q_t = j|\lambda^m)}{P(X_s|\lambda^m)}.$$

- **M-step: Calculate λ^{m+1} that maximizes $G(\lambda, \lambda^m)$**

Following Eq. (2.19), we calculate λ^{m+1} as:

$$\lambda^{m+1} = \arg\max_{\lambda} G(\lambda, \lambda^m)$$
$$= \arg\max_{\pi_i} G(\pi_i, \lambda^m) + \arg\max_{a_{ij}} G(a_{ij}, \lambda^m) + \arg\max_{b_j} G(b_j, \lambda^m). \tag{2.26}$$

We can then maximize $G(\lambda, \lambda^m)$ by maximizing the individual terms separately, subject to probability constraint (assuming discrete distributions):

$$\sum_{i=1}^N \pi_i = 1; \sum_{j=1}^N a_{ij} = 1 \text{ for } \forall i; \text{ and } \sum_{k=1}^K b_j(k) = 1 \text{ for } \forall j,$$

and set the derivative equal to zero:

$$\frac{\partial G(\pi_i, \lambda^m)}{\partial \pi_i} = 0; \frac{\partial G(a_{ij}, \lambda^m)}{\partial a_{ij}} = 0; \text{ and } \frac{\partial G(b_j, \lambda^m)}{\partial b_j} = 0.$$

By using the Lagrange multipliers ψ, a function such that

$$F(x) = \sum_i y_i log(x_i); \quad \text{where} \quad \sum_i x_i = 1, \tag{2.27}$$

and the derivative

$$\frac{\partial}{\partial x_i}\left[\sum_i y_i log(x_i) + \psi(\sum_i x_i - 1)\right] = 0, \tag{2.28}$$

can be proven to achieve maximum value at

$$x_i = \frac{y_i}{\sum_i y_i}. \tag{2.29}$$

Thus, using this formation, we obtain the new estimate of the model parameters $\lambda^{m+1} = (\pi, A, B)$ as:

$$\pi = \frac{\frac{1}{P(X_s|\lambda^m)} P(X_s, q_1 = i|\lambda^m)}{\frac{1}{P(X_s|\lambda^m)} P(X_s|\lambda^m)} = \gamma_1(i), \tag{2.30}$$

$$\begin{aligned} a_{ij} &= \frac{\frac{1}{P(X_s|\lambda^m)} \sum_{t=2}^{T} P(X_s, q_{t-1} = i, q_t = j|\lambda^m)}{\frac{1}{P(X_s|\lambda^m)} \sum_{t=2}^{T} P(X_s, q_{t-1} = i|\lambda^m)} \\ &= \frac{\sum_{t=2}^{T} \xi_{t-1}(i,j)}{\sum_{t=2}^{T} \gamma_{t-1}(i)}, \end{aligned} \tag{2.31}$$

$$\begin{aligned} b_j(x_t) &= \frac{\frac{1}{P(X_s|\lambda^m)} \sum_{t=1}^{T} P(X_s, q_t = j|\lambda^m)\delta(x_t, v_k)}{\frac{1}{P(X_s|\lambda^m)} \sum_{t=1}^{T} P(X_s, q_t = j|\lambda^m)} \\ &= \frac{\sum_{t \in x_t = v_k} \gamma_t(j)}{\sum_{t=1}^{T} \gamma_t(j)}, \end{aligned} \tag{2.32}$$

where

$$\begin{aligned} \delta(x_t, v_k) &= 1 \ \text{ if } \ x_t = v_k \\ &= 0 \ \text{ otherwise.} \end{aligned} \tag{2.33}$$

Further details of the HMM theory may be found in (Rabiner and Juang, 1993; Seymore et al., 1999; Ephraim and Merhav, 2002), and the EM algorithm is described in (Dempster et al., 1977; Baum, 1972; Bilmes, 1998).

2.3 Pattern Recognition for HMM-Based ASR Systems

We have shown in Section 2.1 that pattern recognition may be described in terms of mapping $x \rightarrow y$ from a measurement space X into a decision space Y. In the context of HMM-based speech recognition, we substitute "x" with a "speech signal of spoken utterance" and "y" with "strings of words." Then, the pattern recognition approach is used to map a speech signal into a set of meaningful strings of words.

Mapping a speech signal into meaningful words can be performed using multi-level pattern recognition, since the acoustic speech signals can be structured into a hierarchy of speech units such as subwords (phonemes), words, and strings of words (sentences) (Werner, 2000). A generic automatic speech recognition system, as shown in Figure 2.12, is composed of five components:

1. Feature extraction, which processes the speech signal into a set of observation feature vectors X_s by removing redundant or unimportant information such as the fundamental frequencies or noise,

2. Acoustic model, which estimates the acoustic probabilities $P(X_s|\lambda)$ that the observation feature vectors X_s have been generated by the subword (phoneme) models λ,

3. Pronunciation lexicon, which estimates the word probabilities $P(\lambda|W)$ given the sequence of subwords (phonemes) generated by the model λ,

4. Language model, which estimates the prior probability of sequences of words $P(W)$, and

5. Search algorithm, which determines the most probable string of words $\hat{W}$ among all possible word strings W given the observation feature vector X_s, which can be estimated from the evidence of the acoustic modeling, the lexicon, and the language modeling.

To recognize the speech signal as the correct meaning in words, the ASR system (or more precisely the acoustic model, lexicon, and language model) needs to know the stochastic laws - encoded in the parameters - that govern the mapping. In practical applications, however, these stochastic laws are never explicitly known. The common core is "learning by examples" where all such parameters in the acoustic model, the lexicon and the language model have to be learned from a collection of samples in the training set. The following sections describe each of these five components in turn.

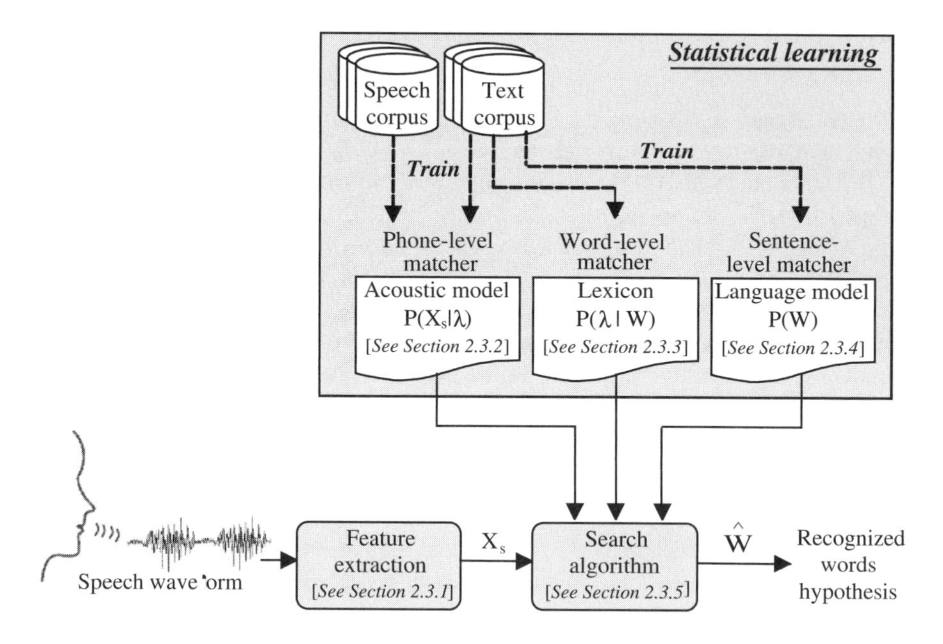

Fig. 2.12. A generic automatic speech recognition system, composed of five components: feature extraction, acoustic model, pronunciation lexicon, language model and search algorithm.

2.3.1 Front-end Feature Extraction

The purposes of the first stage of the speech recognition process are to extract those features that carry as much "important" information as possible about the linguistic content of the speech signal, and to suppress or eliminate "irrelevant" (non-linguistic) features.

Many variants of feature extraction techniques have been developed, e.g., linear prediction coefficients (LPC) (Markel and Jr., 1976), cepstrum analysis (Oppenheim and Schafer, 1975; Fukada et al., 1992), perceptual linear prediction (PLP) (Hermansky, 1990), and modulation-filtered spectrograms (MSG) (Kingsbury, 1998). The most widely used feature extraction technique in speech recognition is based on mel-frequency cepstral coefficients (MFCC) (Seltzer, 1999). It combines the cepstrum analysis with a nonlinear weighting in frequency (filter-bank) (Deng and O'Shaughnessy, 2003).

According to Chen (2004), many speech sounds may be considered as a convolution of two independent components (see Figure 2.13):

- a source $e[n]$, which is the air flow at the vocal chords (excitation).
- a filter $h[n]$, which is the resonance of the vocal tract which changes over time.

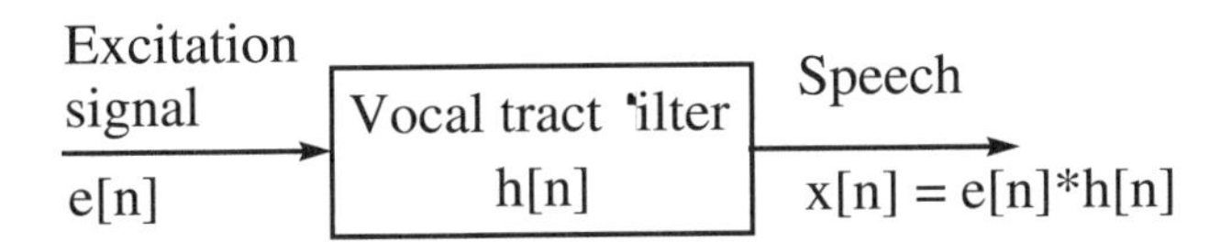

Fig. 2.13. Source-Filter model of the speech signal $x[n] = e[n] * h[n]$.

The linguistic content of the speech signal, i.e., phoneme classification, is mostly dependent on the characteristics of the vocal tract filter. Thus, the technique of cepstral analysis is basically used to extract the desirable information. This requires separate estimation of the individual components; hence a deconvolution of the source and the filter (Deng and O'Shaughnessy, 2003) is necessary. Although such deconvolution is generally nondeterministic, it has some success when applied to speech because the relevant convolved signals forming speech have very different time-frequency behavior (Deng and O'Shaughnessy, 2003). In principle, cepstral deconvolution transforms a convolution into a sum of two signals using a discrete Fourier transform (DFT), a logarithm function, and an inverse DFT (IDFT), as illustrated in Figure 2.14.

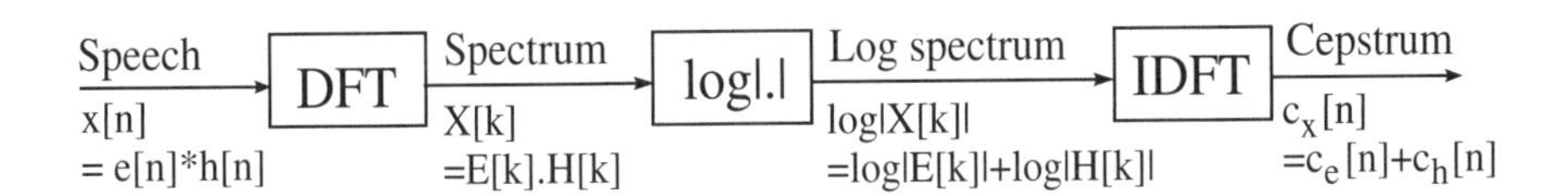

Fig. 2.14. Source-filter separation by cepstral analysis.

Figure 2.15(a) shows an example of a windowed speech waveform. After the DFT is performed, the two convolved components have multiplicative correlates in the speech spectrum, where the "quickly varying" part represents the excitation signals of the sources, and the "slowly varying" part (spectral envelope) represents the vocal tract filter (see Figure 2.15(b)). Once the logarithm of the spectral magnitude is taken, the multiplicative relation between the excitation signal and the spectral envelope are transformed into an additive relation. Then, by performing the IDFT on logarithm of spectral magnitude, one separates the slowly varying part from the quickly varying part. The slowly varying part results the cepstral components at low quefrency, while the quickly varying part results the cepstral components at high quefrency (see Figure 2.15(c)). Finally, the goal is to capture only the low-quefrency part of the cepstrum which is the gross spectral envelope of the input speech signal (see Figure 2.15(d)), and therefore represents an approximation of the vocal tract transfer function.

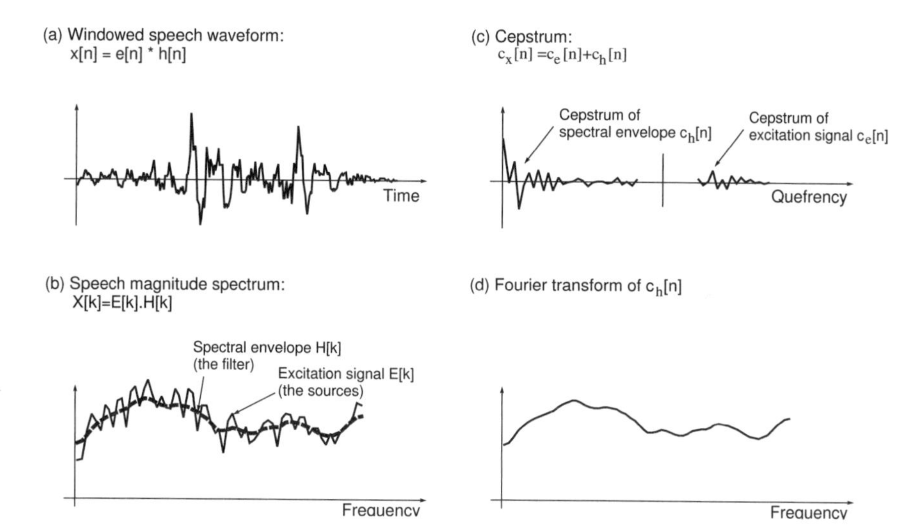

Fig. 2.15. (a) A windowed speech waveform. (b) The spectrum of Figure 2.15(a). (c) The resulting cepstrum. (d) The Fourier transform of the low-quefrency component.

The detailed processing steps for MFCC are illustrated in Figure 2.16 and described as follows:

1. **Spectral shaping**

   This is the process of converting the analogue speech signal into a digital signal or A/D conversion (Rosell, 2006), which consists as follows:

   - Filtering (Pre-processing)
     First, the analog speech signal is passed through a low-pass filter with a cutoff frequency f_m typically in the region of 5.6 kHz, to reduce the high-frequency noise. This seems reasonable since the frequency range of the human voice does not extend much beyond 5 kHz (Werner, 2000).

   - Digitization (A/D Conversion)
     The next step is analog to digital conversion, known as sampling and quantization. Usually, speech is recorded or sampled at a sampling rate f_s of 16 kHz over a microphone or 8 kHz over a standard telephone (Werner, 2000). This over-sampling satisfies the Nyquist criterion, $f_s \geq 2f_m$, to avoid the aliasing phenomenon. For a 16-kHz sampling rate and 16-bit quantization, the bit rate equals 256 kbits/s.

- Preemphasis
  The preemphasis filter is intended to boost the high frequency of the signal spectrum approximately 20 dB per decade (an order of magnitude increment in frequency), since voiced sections of speech signal naturally have an attenuation of approximately 20 dB per decade due to physiological characteristics of the speech production system (Picone, 1993; Jurafsky, 2007).

- Framing
  The sampled waveform is then divided into successive overlapping frames. Although speech is non-stationary through the movement of the articulations, overall it can be considered quasi-stationary (Bourlard and Morgan, 1994). This means, over a short period of time, the statistics of the speech signal do not differ significantly from sample to sample (piecewise stationary process). Most ASR systems measure the feature vector over segments, called *frames*, of around 16-32 ms and are updated every 8-16 ms (Kingsbury, 1998). The frames are taken at distances less than a frame length. By introducing overlapping, the transitions between frames can be smoothed (Maekinen, 2000). In our case, a frame length of 20 ms and a frame shift of 10 ms are used.

- Windowing
  Each frame is then fed to a Hamming window (Harris, 1978) to enhance the harmonics and to eliminate discontinuities at the edges for subsequent Fourier transforms. A common Hamming window, which was used in this work, is defined as

$$w[n] = 0.54 + 0.46 \cos\left(\frac{2\pi n}{N-1}\right). \tag{2.34}$$

2. **Spectral Analysis**

The spectral analysis processes the spectrum in order to capture the important aspects of the signal using a deconvolution technique as described earlier. The process include:

- Discrete Fourier transform (DFT)
  The standard algorithm to compute DFT is the fast Fourier transform (FFT) (Brenner and Rader, 1976). It is applied to the windowed frames and compute a short time power spectrum.

$$\begin{aligned} X[k] &= \sum_{n=0}^{N-1} x[n] w[n] e^{-j2\pi kn/N}, \\ P[k] &= |X[k]|^2 = \mathrm{Re}^2\left(X[k]\right) + \mathrm{Im}^2\left(X[k]\right), \end{aligned} \tag{2.35}$$

where n denotes the sample time, and $w[n]$ denotes the Hamming window frame, k denotes the discrete frequency $f(k) = \frac{kf_s}{N}$, f_s denotes the sampling frequency, and N is the length of the FFT .

- Filter bank
  A critical-band-like spectrum is derived by multiplying the power spectrum with a bank of overlapping triangular weighting filters using the mel scale, and integrating the result

$$\tilde{P}[l] = \sum_{k=0}^{K-1} P[k]M_l[k], \quad \text{for} \quad l = 1, 2, \ldots, L, \tag{2.36}$$

  where L is the total number of triangular filter banks $M[k]$, and a scale is used to define the spacing of the filter bandwidths. That is, filters at higher frequencies have wider bandwidths than at lower frequencies. Using the mel scale, the spacing is close to linear for the center frequency f_c below about 1 kHz, and is close to logarithmic for f_c above 1 kHz. The analytical expression defining 1 mel scale is

$$mel(f_c) = 2595 \cdot log(1 + \frac{f_c}{700}). \tag{2.37}$$

  This is done since human hearing is not equally sensitive to all frequencies (less sensitive to higher frequencies).

- Log spectrum computation and inverse DFT (IDFT)
  The dynamic range of the critical-band-like spectrum is compressed when applying the logarithm. Then, cepstral coefficients are computed by applying IDFT. Since the log power spectrum is real and symmetric, the IDFT reduces to a discrete cosine transform (DCT). (Ahmed et al., 1974):

$$c_n = c[n] = \sum_{l=0}^{L-1} \ln \tilde{P}[l]. \cos\left(\frac{\pi n}{2L}(2l+1)\right), \tag{2.38}$$

  where c_n is the nth component of the cepstral coefficients, for $n = 0, \ldots, N-1$.

3. **Parametric transform**

The parametric transform codes the measurements achieved through spectral analysis, and produces a feature vector, as follows:

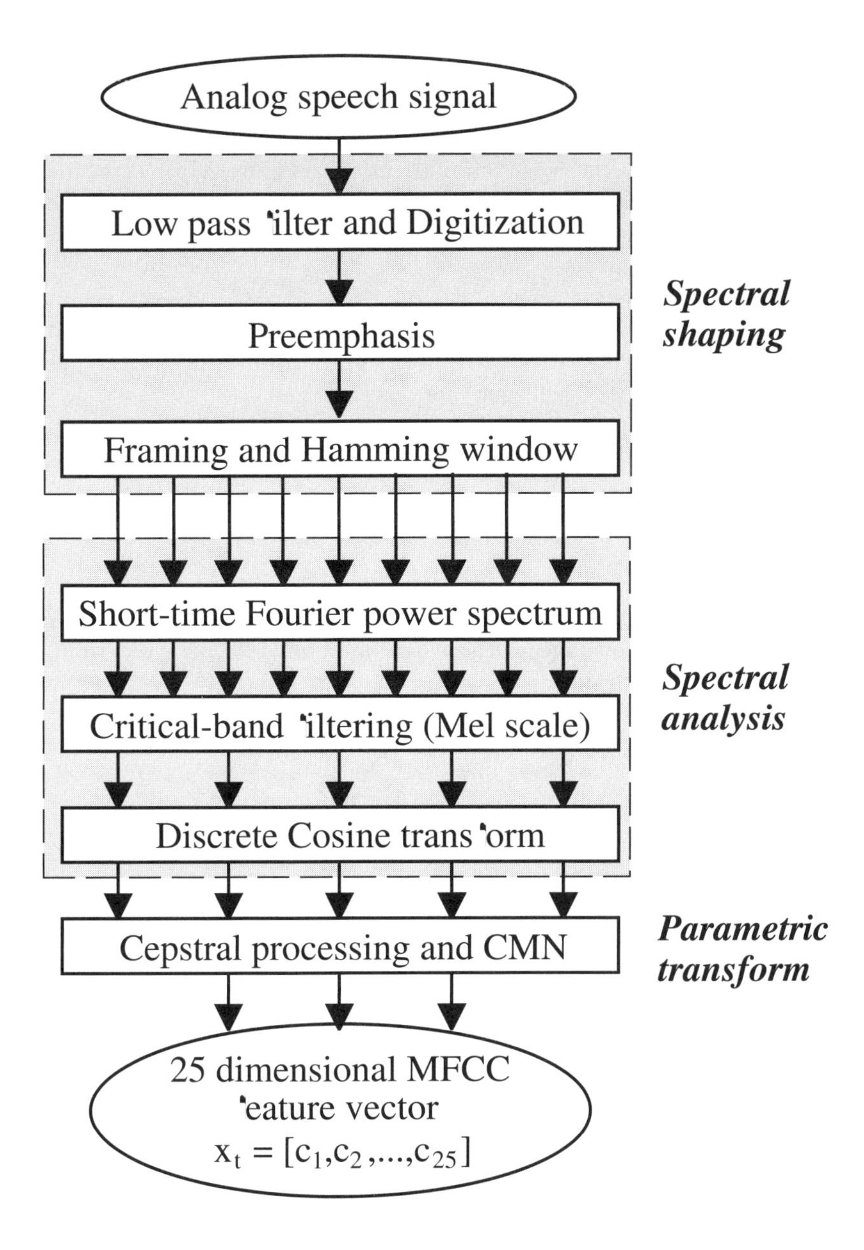

Fig. 2.16. MFCC feature extraction technique, which generates a 25-dimensional feature vector x_t for each frame.

- Low order cepstral coefficients
  As described previously, the envelope of the vocal tract filter changes slowly, and thus correspond to lower order of cepstral coefficients, while the periodic excitation signal (the sources) are at higher order cepstral coefficients (see the cepstrum illustration in Figure 2.15). Thus, in general, only the low order of the cepstral coefficients are used (typically only the first 12 cepstral coefficients).

- Energy coefficient
  The coefficient c_0 is usually not used (Maekinen, 2000), and it is sometimes replaced by the logarithm of the total intensity of the vector, called the energy coefficient (log power):

$$c_E = \ln \sum_{l=0}^{L-1} \tilde{P}[l]. \tag{2.39}$$

- Cepstral mean subtraction (CMS)
  In a way of normalizing for channel effects, a CMS technique is applied, where a time average of the cepstral values is subtracted from the values at each frame :

$$\begin{aligned} \bar{c}_n &= \frac{1}{T} \sum_{T} c_n, \\ c_n^{new} &= c_n - \bar{c}_n. \end{aligned} \tag{2.40}$$

  This is computed over the length T of a complete utterance.

- Dynamic features
  The features described so far have not captured the dynamics of the spectral changes (the slopes). Thus, first and second-order dynamic features (delta and delta-delta coefficients) are usually appended also to the acoustic vector (Furui, 1986).

Based on our experimental measurements, the optimal MFCC feature vector is composed of a 12-order MFCC, 12-order Δ MFCC, and Δ log power, resulting in a total of 25 components in total: $x_t = [c_1, c_2, \cdots c_{25}]$.

In summary, the spectrum is captured over frames, then the spectral properties are analyzed, producing a feature vector which correspond to one point in a multi-dimensional space (see Figure 2.17).

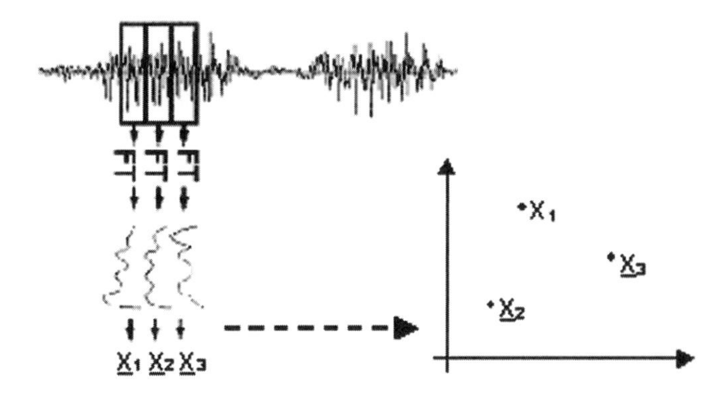

Fig. 2.17. A summary of feature extraction process, producing a feature vector which correspond to one point in a multi-dimensional space.

2.3.2 HMM-Based Acoustic Model

The function of an acoustic model (AM) is to provide the probability of the observation feature vectors X_s, which have been generated by the models λ. HMMs are typically employed as representations of acoustic models λ, where:

1. the short-term speech spectral characteristics are modeled with HMM state distribution, while
2. the temporal speech characteristics are governed by HMM state transitions.

After feature extraction, every frame t of the speech signal (every 10 ms) is represented by a feature vector $x_t = [c_1, c_2, \cdots c_{25}]$ in a multi-dimensional continuous space. In our system, the HMMs perform a state transition every frame, that is, every 10 ms, according to the transition probabilities $a_{ij} = P(q_t = j|q_{t-1} = i)$. Then the state q_j of the process emits a symbol, which is in our case a feature vector x_t, according to a certain emission probability of state j for the vector x_t, $b_j(x_t) = P(X = x_t|q_t = j)$.

The HMMs used to represent the acoustic speech model are in accordance with the Bakis model (see Section 2.5). The state index of the HMM Bakis model increases or remains unchanged as the time increases, leading to a move from left to right on the Markov chain. This traduces the causality of the speech production process:

$$i > j \rightarrow a_{ij} = 0. \tag{2.41}$$

Several types of HMMs may be applied, depending on whether the HMM state distribution observes a certain feature vector or symbol and the type of speech units it represents. The following sections describe the observation density and the speech units in more detail.

Observation Density

The HMM state distribution may use either a discrete or a continuous density to observe a certain feature vector or symbol.

1. Discrete observation feature vector

   The discrete approach considers the case where the observation sequence of each state j belongs to a finite set V of K possible observations $V_j = V_{j1}, V_{j2}, ..., V_{jK}$ as outlined in Figure 2.18.

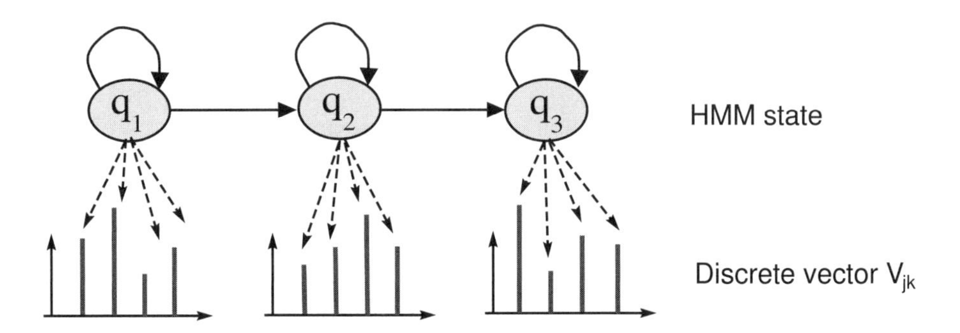

Fig. 2.18. Discrete HMM observation density where the emission statistics or HMM state output probabilities are represented by discrete symbols.

In this case, the HMM state output probability is defined as

$$\begin{aligned} p(x_t|q_j) &= P(X = x_t|Q = q_j) \\ &= P(x_t = V_{jk}|q_t = j) \end{aligned} \tag{2.42}$$

Since a feature vector represents a point in a multi-dimensional continuous space, the number of different possible vectors are infinite. Therefore, a vector quantization (VQ) (Billi, 1982) technique is usually applied to quantize the input to the system.

Although this approach significantly reduces computational speed, there is obviously significant reduction in the dimensionality of the original continuous data, at the cost of a certain loss of signal information. Indeed, at least for some applications, there might be a serious degradation associated with such a discretization of the continuous signal.

2. Continuous observation feature vector

   A typical use in LVCSR is the continuous Gaussian mixture model (GMM) approach. Here, instead of partitioning the space into discrete clusters, the continuous observation space is modeled using Gaussian multivariate densities, which are in turn weighted and added to compute the emission likelihoods of each of the states or the state output probability.

   The Gaussian components are state-specific (see Figure 2.19) and parameterized by the mean vector (representing the mean of the component as a d-dimensional vector) and by the covariance matrix (describing the metric of the space spanned by d-dimension). The training procedure is based on the EM algorithm which corresponds to the training method of the HMM described in section 2.2.3.

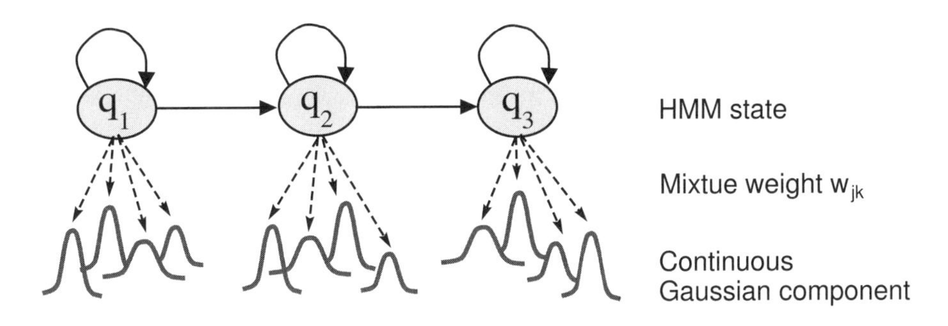

Fig. 2.19. Continuous GMM, where the continuous observation space is modeled using mixture Gaussians (state-specific). They are weighted and added to compute the emission statistic likelihoods (HMM state output probabilities).

The HMM state output probability, $p(x_t|q_j)$, is usually calculated from the state PDF, $P(X|Q)$, as

$$\begin{aligned} p(x_t|q_j) &= P(X = x_t|Q = q_j) \\ &= \sum_{k=1}^{K} w_{jk}\mathcal{N}(x_t; \mu_{jk}, \Sigma_{jk}), \end{aligned} \tag{2.43}$$

where w_{jk} is the mixture weight for the k_{th} mixture in state q_j, and $\mathcal{N}(.)$ is a Gaussian function with mean vector μ_{jk} and covariance matrix Σ_{jk}. The HMM segmental likelihood, $P(X_s|\lambda)$, is then calculated from the joint probability of observation and the state sequence, taken over all state sequences (total likelihood) or approximately over just the most likely state sequence (Viterbi path) (Holmes and Huckvale, 1994).

Speech Unit Representation

As an acoustic model, an HMM may be associated with any temporal property of the speech unit, such as words, syllables or sub-words (phonemes). In the word-unit model, an HMM is associated with each word, whereas in the phoneme-unit model the HMM is associated with each phoneme.

A good acoustic model is able to accurately capture coarticulation effects; i.e., the acoustic and articulatory variability that arises when the articulatory patterns of neighboring speech segments overlap. This represents a fundamental part of language sound systems that allows for dynamic transitions between adjacent phoneme segments (both within and across words) (Scarborough, 2004). As a result, phonemes may have very different waveforms when produced in the context of other phonemes (Rabiner and Juang, 1993). Therefore, wider unit models, allow for a better capturing of the coarticulation effects (Pfau et al., 1997).

However, as described in Section 1.4.1, word-unit and syllable-unit models are impractical for LVCSR systems due to the large amount of training data needed, the large decoding search space, and the inefficiency of expanding the vocabulary system. The phoneme units are thus a natural choice since there are only a few of them and their frequency of appearance in the training data is much higher. Figure 2.20 illustrates an example of a monophone acoustic model. A standard solution to the coarticulation problem is to extend the phoneme units to include context (Smith et al., 2001).

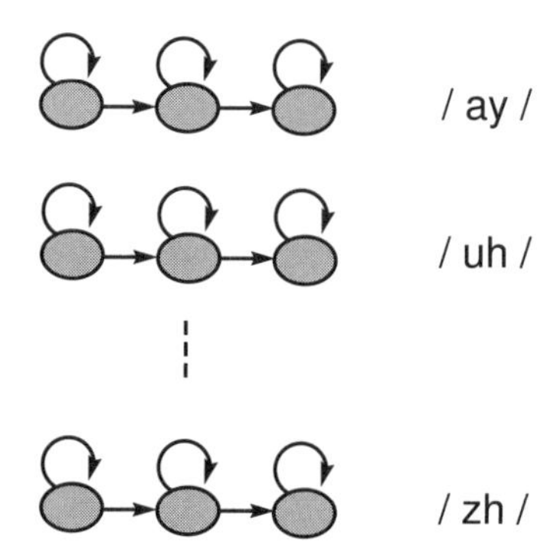

Fig. 2.20. Structure example of the monophone /a/ HMM acoustic model.

Most of the current LVCSR systems use the context-dependent triphone as the fundamental acoustic unit. Context-dependent triphone units (see Figure 2.21) have the same structure as context-independent phonetic (monophone) units, but are trained on data with immediately preceding and following phonetic contextual information (O'Neill et al., 1998).

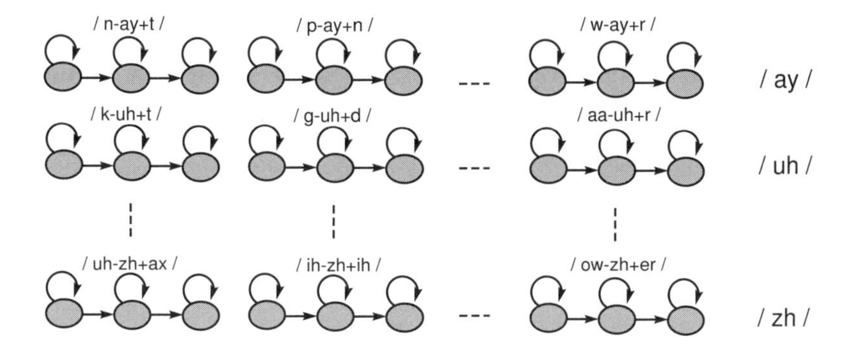

Fig. 2.21. Structure example of the triphone $/a^-, a, a^+/$ HMM acoustic model.

Parameter Tying

If the amount of training data is not sufficient to obtain a reliable estimate of the model parameters, the overall performance may degrade significantly. It seems therefore necessary to reduce the model size. The most common solution is to share some of the model parameters by tying the state output probability distributions among different HMMs as illustrated in Figure 2.22.

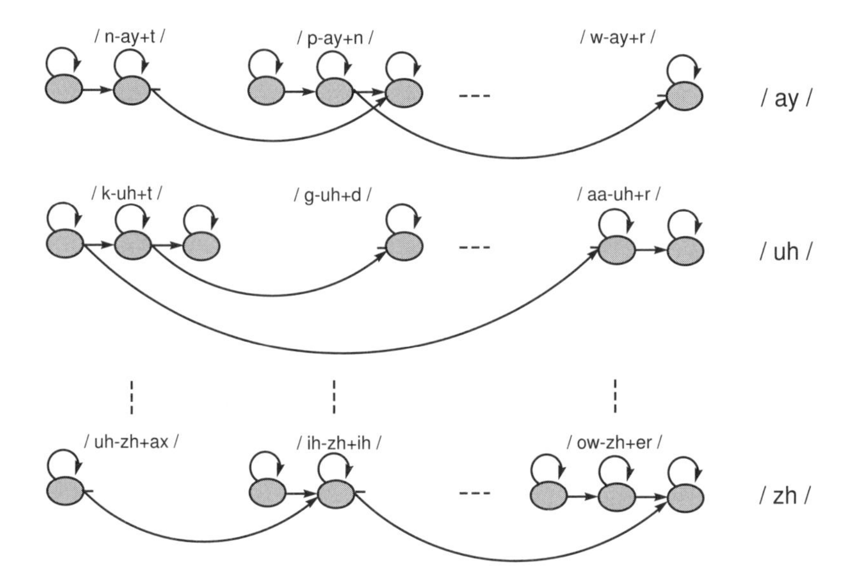

Fig. 2.22. Shared-state structures of the triphone $/a^-, a, a^+/$ HMM acoustic model.

There are two major methods to create shared-state structures, i.e., the decision tree clustering (Sagayama, 1989; Lee et al., 1990; Young et al., 1994; Zhao et al., 1999) and the successive state splitting (SSS) algorithms (Takami and Sagayama, 1992; Girardi, 2001; Jitsuhiro et al., 2004).

Figure 2.23 shows an example of the phonetic decision tree for HMM state of triphone with the central phoneme /*ay*/. By asking questions, the states which have similar acoustic contexts enter into the same leaf node and are tied together. There are acoustically similar, hence it make sense to allow them share the same distribution (Zhao et al., 1999).

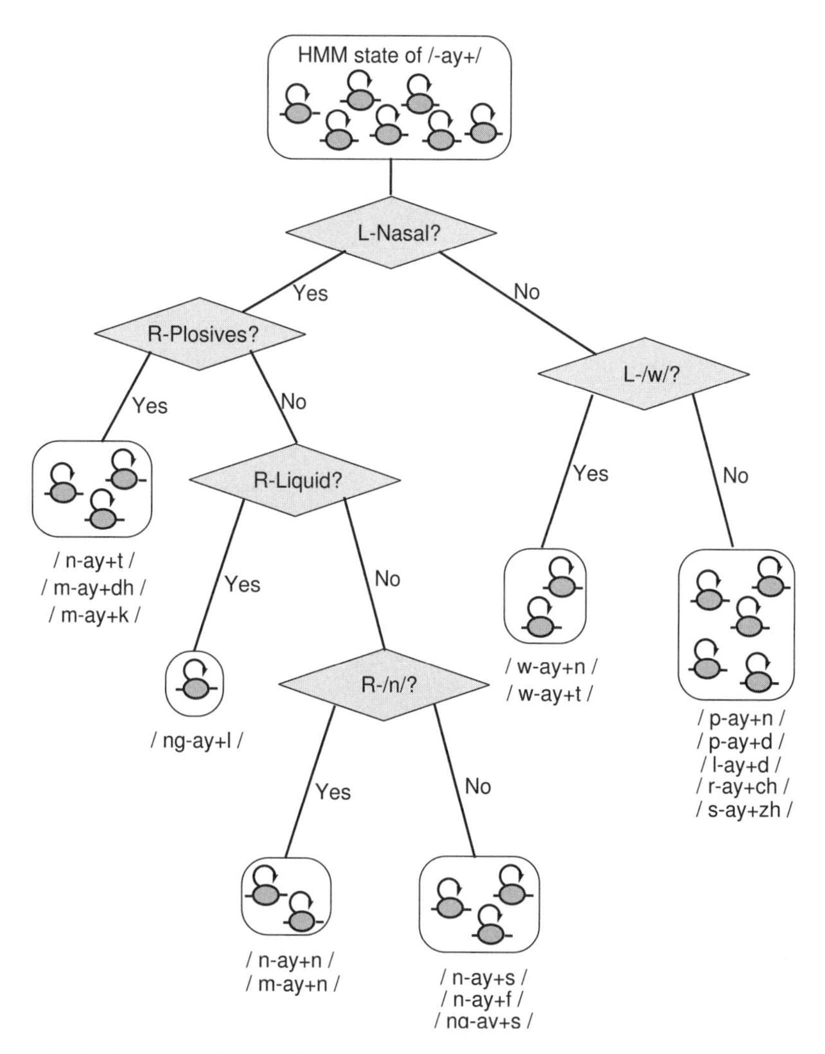

Fig. 2.23. An example of a phonetic decision tree for HMM state of the triphone with the central phoneme /*ay*/.

The phonetic decision tree clustering can only create contextual variations, while in contrast the SSS algorithm can create both contextual and temporal variations (see Figure 2.24). This method use the maximum likelihood (ML) criterion to choose a model. However, owing to the nature of the ML estimation, the likelihood value for training data increases as the number of parameters increases. In our case, the SSS algorithm used here was based on the minimum description length (MDL) optimization criterion. Details on the MDL-SSS can be found elsewhere (Jitsuhiro et al., 2004).

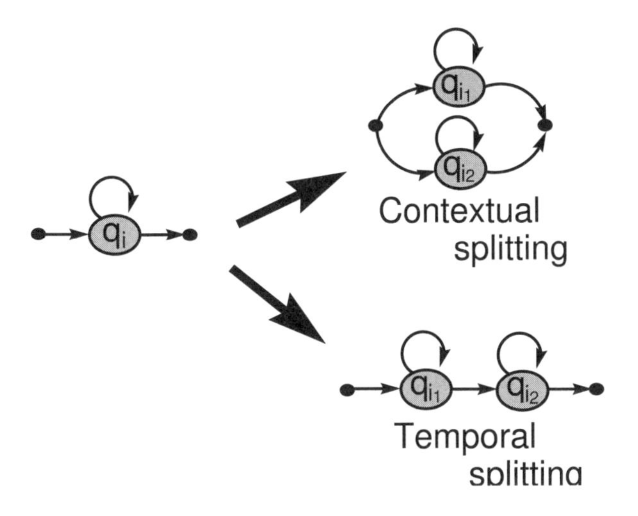

Fig. 2.24. Contextual splitting and temporal splitting of SSS algorithm (After Jitsuhiro, 2005).

2.3.3 Pronunciation Lexicon

The lexicon describes the pronunciation of all words in the vocabulary, which is achieved through in the implementation of a lexical tree-based search (Fetter, 1998). An example of a lexicon tree using sub-word (phoneme) units is illustrated in Figure 2.25. As can be seen, the lexical tree contains all of the words along with their pronunciation dictionary (often there are multiple pronunciations of a word). Here, each node may be associated with a subword (phoneme) and be shared by multiple words with the same partial pronunciation. A terminal node of the lexical tree signifies a unique word.

In this way, we can estimate the word probabilities $P(\lambda|W)$ given the sequence of subwords (phonemes) generated by the HMM model λ.

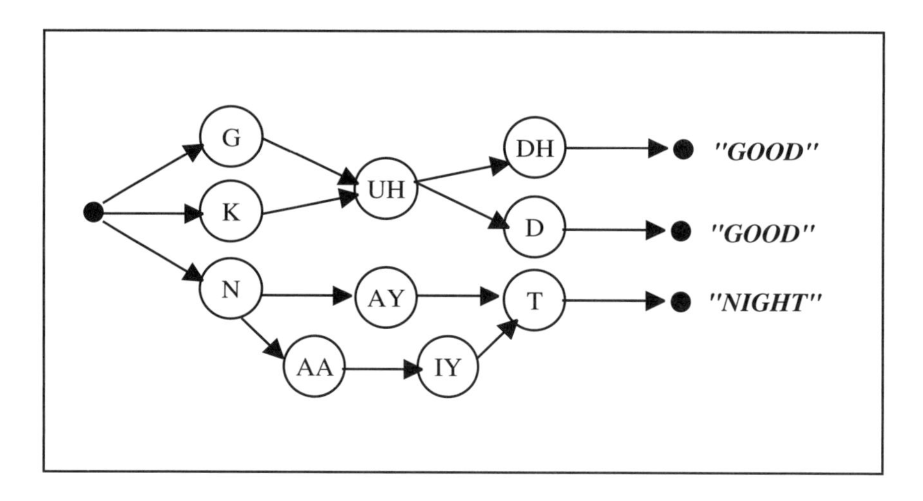

Fig. 2.25. Example of a tree-based pronunciation lexicon.

2.3.4 Language Model

The function of a language model (LM) is to provide the speech recognizer with the a priori probability $P(W)$ of a hypothesized word sequence $W = w_1, w_2, ..., w_n$ (Fetter, 1998). This tells us, whether we heard "bad boy" or "pad boy." $P(W)$ can be decomposed into the product of the word probabilities that make up the word sequence W:

$$\begin{aligned} P(W) &= P(w_1)P(w_2|w_1)P(w_3|w_1w_2)\cdots P(w_n|w_1w_2\ldots w_{n-1}) \\ &= \prod_{i=1}^{n} P(w_i|w_1\ldots w_{i-1}). \end{aligned} \tag{2.44}$$

However, such expressions cannot lead to reliable probability estimates, given the arbitrarily long word histories that would require enormous amounts of training data. Fortunately, the formula can be approximated using an N-gram model. One popular and effective approach is the bigram or trigram LM, which assumes that the probability of any given word is determined by the N-previous word strings. The bigram LM provides a probability of a complete word string W, given a one word history:

$$\begin{aligned} P(W) &= P(w_1)P(w_2|w_1)P(w_3|w_2)\cdots P(w_n|w_{n-1}) \\ &= \prod_{i=1}^{n} P(w_i|w_{i-1}). \end{aligned} \tag{2.45}$$

Furthermore, the trigram LM provides a probability of a complete word string W, given a two-word history:

$$\begin{aligned} P(W) &= P(w_1)P(w_2|w_1)P(w_3|w_2,w_1)\cdots P(w_n|w_{n-1},w_{n-2}) \\ &= \prod_{i=1}^{n} P(w_i|w_{i-1},w_{i-2}). \end{aligned} \tag{2.46}$$

Typically, the LM grammar does not need to contain probabilities for all possible word pairs, but only for the most frequently occurring N-grams. Then, it uses a back-off mechanism to fall back on unigram probability if the desired N-gram is not found. Further details are provided in (Fetter, 1998; Russel and Norvig, 1995).

2.3.5 Search Algorithm

The search algorithm of a statistical framework for speech recognition problems is to choose the most probable string of words $\hat{W}$ among all possible word strings W, given the observation feature vector X_s, the so-called maximum *a posteriori* (MAP) (DeGroot, 1970; Sorenson, 1980):

$$\hat{W} = \arg\max_{W} P(W|X_s). \tag{2.47}$$

Using the Bayes rule, we have:

$$\begin{aligned} \hat{W} &= \arg\max_{W} \frac{P(X_s|W)P(W)}{P(X_s)} \\ &= \arg\max_{W} \frac{P(X_s|\lambda)P(\lambda|W)P(W)}{P(X_s)}. \end{aligned} \tag{2.48}$$

Since the probability of the sequence utterances $P(X_s)$ is constant for all words during recognition, it can be ignored, resulting in

$$\hat{W} = \arg\max_{W} P(X_s|\lambda)P(\lambda|W)P(W), \tag{2.49}$$

where

$P(W|X_s)$ is the global posterior probability of the word string W given the observation feature vector X_s,

$P(X_s|W)$ is the global likelihood that the observation feature vector X_s was produced by the word string W,

$P(X_s|\lambda)$ is the likelihood that the observation feature vector X_s was generated by the model λ, which is provided by an acoustic model,

$P(\lambda|W)$ is the word probabilities given the sequence of subwords (phonemes), which is provided by the pronunciation lexicon,

$P(W)$ is the a priori probability of the word string W, which is provided by the language model, and

$P(X_s)$ is the probability of the sequence X_s.

The multi-level probability estimation of this statistical framework is outlined in Figure 2.26.

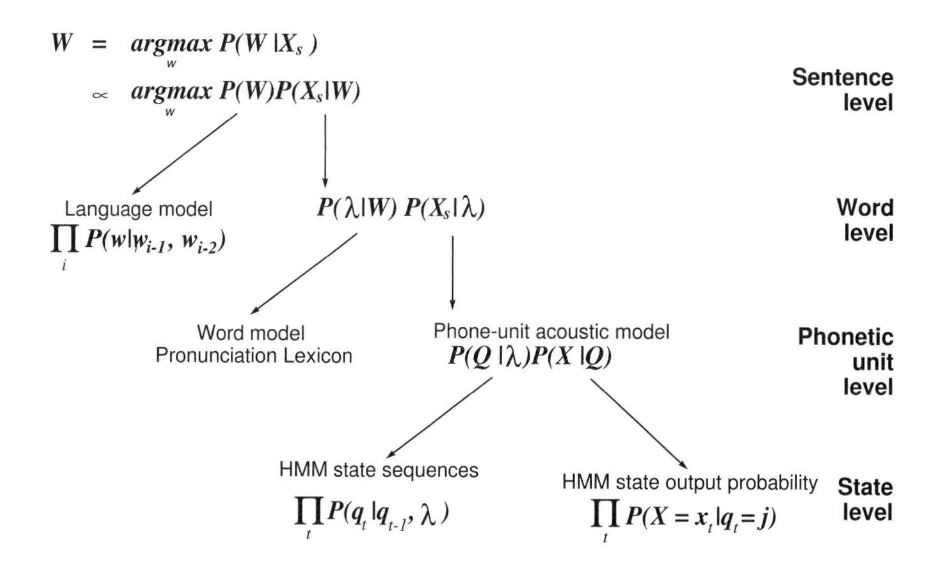

Fig. 2.26. Multi-level probability estimation of statistical ASR.

One problem for the search algorithm is to determine at which point a speaker starts and finishes an utterance. While the recognizer allows users to speak naturally (the word boundaries are not well known), the machine needs to determine the content and utilize special methods to determine utterance boundaries. An ASR system may perform continuous speech recognition offline or online. In offline recognition, the recognition takes places at the end of the sequence of words. In online recognition, the system performs recognition every time it detects a pause in the speech production. In both cases, the goal is to find the most probable sequence of the words spoken from among all possible sequences. Different techniques of decoding or search algorithms have been proposed. The most often used approach, however, is the Viterbi algorithm, which corresponds to the decoding method of the HMM Problem (see Section 2.2.3 for more details).

3

Graphical Framework to Incorporate Knowledge Sources

In this chapter, we introduce the design of our proposed framework, the so called GFIKS (graphical framework to incorporate additional knowledge sources). It is based on a graphical model representation that makes use of additional knowledge sources in a statistical model as shown in Figure 3.1. This approach is meant to be broadly useful in the sense that it can be applied to many existing modeling problems with their respective model-based likelihood functions.

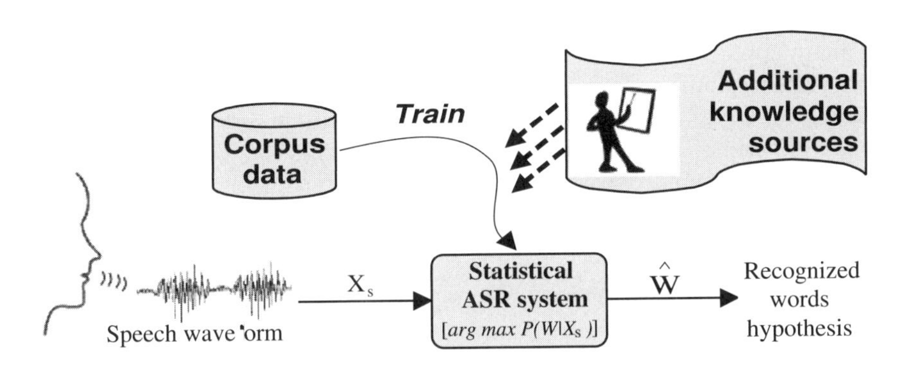

Fig. 3.1. Incorporating knowledge into corpus-based statistical ASR system.

In Section 3.1, we review graphical model representation, including probability theory and graph theory. In Section 3.2, we introduce GFIKS's procedure for knowledge incorporation, including how to define the causal relationships between information sources (Section 3.2.1), how to do direct inference (Section 3.2.2), and how to proceed when direct inference is intractable (Sections 3.2.3 and 3.2.4). In Section 3.3, we discuss the general issues and possibilities of incorporating knowledge sources in the statistical ASR system,

S. Sakti et al., *Incorporating Knowledge Sources into Statistical Speech Recognition*,
Lecture Notes in Electrical Engineering 42, DOI: 10.1007/978-0-387-85830-2_3,

such as what type of knowledge sources are important (Section 3.3.1) and at which level of ASR they should be incorporated (Section 3.3.2).

3.1 Graphical Model Representation

Graphical models, a marriage between probability theory and graph theory, provide a natural tool for dealing with two problems: uncertainty and complexity. Fundamental to the idea of a graphical model is the notion of modularity: a complex system is built by combining simpler parts. Probability theory serves as the glue whereby the parts are combined, ensuring that the system as a whole is consistent and providing ways to interface models to data. Graph theory provides both an intuitively appealing interface by which humans can model highly interacting sets of variables and a data structure that lends itself naturally to the design of efficient general-purpose algorithms (Jordan, 1999).

In the following, we first review the probability theory in Section 3.1.1 and then describe in Section 3.1.2 how a graphical model can compactly represent a joint probability distribution.

3.1.1 Probability Theory

Over the last decade, the Bayesian approach has become increasingly popular in many application areas. This statistical method provides a paradigm for both statistical inference and decision making under conditions of uncertainty (Bernardo, 2001). It is based on a probabilistic framework that encodes our beliefs or actions in situations of uncertainty. Information from several models may also be combined based on the Bayesian framework to achieve better inference and to better account for modeling uncertainty.

Learning a Probability

The basic axioms in classical probability calculus (Kjaerulff and Madsen, 2005) are described as follows:

1. Normality

   A probability of an event a, $P(a)$, is a number in the unit interval $[0, 1]$; a non-negative real number less than or equal to 1.

$$0 \leq P(a) \leq 1.$$

2. Certainty

   A probability equals 1 if and only if the associated event has happened for sure:

$$P(a) = 1 \text{ if and only if } a \text{ is certain.}$$

3. Additivity

   If two events cannot co-occur, i.e. they are mutually exclusive, then the probability that either one of them occurs equals the sum of the probabilities of their individual occurrences:

$$P(a \;\; or \;\; b) \equiv P(a \vee b) = P(a) + P(b).$$

4. Conditional probability

   The probability of the co-occurrence of two events, a and b, may be computed as the product of the probability of event a occurring conditionally on the fact that event b has already occurred and the probability of event b occurring. It may also be computed conversely, as the product of the probability of event b occurring conditionally on the fact that event a has already occurred and the probability of event a occurring.

$$P(a \;\; and \;\; b) \equiv P(a \wedge b) \equiv P(a, b) = P(a|b)P(b) = P(b|a)P(a),$$

   where

   $P(a, b)$ is called the joint probability of the events a and b.
   $P(a|b)$ is called the conditional probability of event a given the occurrence of b.

   This *conditional probability* is the basic concept in the Bayesian treatment of certainties. Given event b, the conditional probability of event a is x, written as

$$P(a|b) = x.$$

   This means that if event b is true and everything else known is irrelevant for event a, then the probability of event a is x. This conditional probability $P(a|b)$ can be obtained from the joint probability $P(a, b)$ through *conditional normalization* with respect to $P(b)$:

$$P(a|b) = \frac{P(a, b)}{P(b)}.$$

Bayes's Rule

The fundamental rule of probability calculus, based on Axiom 4 described in the previous section, can be written as

$$P(a, b) = P(a|b)P(b) = P(b|a)P(a). \tag{3.1}$$

Bayes's rule follows immediately:

$$P(b|a) = \frac{P(a|b)P(b)}{P(a)}. \tag{3.2}$$

The prior distribution, $P(b)$, expresses our initial belief about b, and the posterior distribution, $P(b|a)$, expresses our revised belief about b after obtaining a. The quantity $P(a|b) = L(b|a)$ is called the likelihood for a given b, since it is a measure of how likely that for a, b is the cause. Bayes's rule then tells us how to obtain the posterior distribution by multiplying the prior $P(b)$ by the ratio $P(a|b)/P(a)$. In general (Kjaerulff and Madsen, 2005),

$$\text{posterior} \;\propto\; \text{likelihood x prior knowledge.}$$

In training a model from data based on the principle of pattern recognition, we can consider a prior distribution, $P(M)$, for a random model variable M, as expressing a set of possible models. For any value D, expressing data, the quantity $P(D|M)$ is the likelihood function for M given data D. The posterior distribution for M given data D is then

$$P(M|D) \;\propto\; P(D|M)P(M),$$

which provides a set of goodness-of-fit measures for models M. Having specified $P(D|M)$ and $P(M)$, the mechanism of the theorem provides a solution to the problem of how to learn from data (Bernardo and Smith, 1994).

Chain Rule of the Probability Product

In general, for a PDF, $P(Z)$, over a set of random variables $Z = (Z_1, Z_2, ..., Z_K)$, can decompose it into the product of conditional probability distributions by using the chain rule:

$$\begin{aligned} P(Z_1, Z_2, ..., Z_K) &= P(Z_1|Z_2, ..., Z_K)P(Z_2, ..., Z_K) \\ &= P(Z_1|Z_2, ..., Z_K)P(Z_2|Z_3, ..., Z_K)...P(Z_{K-1}|Z_K)P(Z_K) \\ &= \prod_{k=1}^{K} P(Z_k|Z_{k+1}, ..., Z_K) \end{aligned} \tag{3.3}$$

It should be noted that the actual conditional distributions that comprise the factors of the decomposition are determined by the order in which we select the head variables of the conditional distributions. Thus, there are K different factorizations of $P(Z_1, Z_2, ..., Z_K)$; accordingly, they represent no independence statement (Kjaerulff and Madsen, 2005).

3.1.2 Graphical Model

Graph theory provides an excellent language for communicating and discussing dependence and independence relations among problem-domain variables. Consequently, it gives us a very intuitive language for representing such dependence and independence statements (Jensen, 1998).

There are two major kinds of graphical models: those based on *undirected* graphs and those based on *directed* graphs. Our main focus is on directed graphs, the so-called Bayesian network (BN) (Jensen, 2001; Heckerman, 1995; Murphy, 2001).

Bayesian networks, also known as probabilistic networks, are graphical models of causal interactions among a set of variables, where the variables are represented as nodes (also known as vertices) of a graph and the interactions (direct dependencies) as directed links (also known as arcs and edges) between the nodes. Any pair of unconnected/nonadjacent nodes of such a graph indicates conditional independence between the variables represented by these nodes under particular circumstances, which can be easily read from the graph (Kjaerulff and Madsen, 2005). Therefore, probabilistic networks capture a set of conditional dependence and independence properties associated with the variables represented in the network. The representation of those probability properties in the Bayesian network is described in the following sections.

Bayes's Rule Through Arc Reversal

The application of Bayes's rule can also be given a graphical interpretation. Consider, for example, two variables a and b and a model $P(a,b) = P(b|a)P(a)$. Again, following the discussion in Section 3.1.2, this model can be represented graphically as indicated in Figure 3.2(a).

To apply Bayes's rule on this model, we perform the following calculations:

1. $P(a,b) = P(b|a)P(a)$,
2. $P(b) = \sum_a P(a,b)$, and
3. $P(a|b) = \frac{P(a,b)}{P(b)}$,

whereby we obtain the equivalent model shown in Figure 3.2(b). Thus, one way of interpreting the application of Bayes's rule is through the so-called *arc reversal*. The work in (Shachter, 1990) has exploited this approach in its proposed arc reversal algorithm for inference in probabilistic networks.

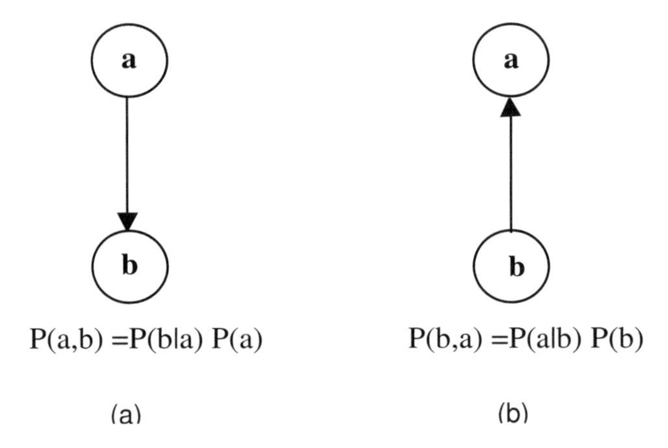

Fig. 3.2. Two equivalent models that can be obtained from each other through arc reversal of Bayes's rule, since P(a,b)=P(b,a).

Conditional Probability

The conditional probability distributions of probabilistic networks are of the form $P(a|b)$, where a is a single variable and b is a (possibly empty) set of variables. a and b are sometimes called the head and the tail, respectively, of $P(a|b)$.

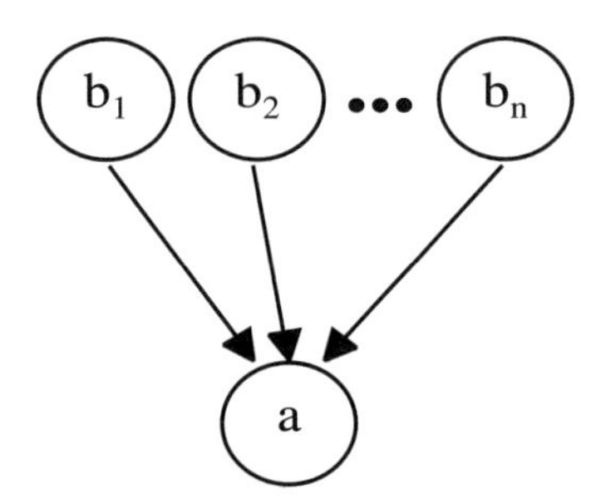

Fig. 3.3. Graphical representation of $P(a|b_1, b_2, ..., b_n)$.

The relation between a and $b = b_1, b_2, ..., b_n$ can be represented graphically as a DAG (directed acyclic graph) illustrated in Figure 3.3, where the child node is labeled a and the parent nodes are labeled $b_1, b_2, ..., b_n$. The set of parents and children of these nodes can be denoted by $Pa(b)$ and $Ch(a)$, respectively.

In addition to the structure, it is also necessary to specify the parameters of the model. In our work, we use a circle node to denote a continuous variable and a square node to denote a discrete variable. The variable values at each node are specified by a conditional probability distribution (CPD). This can be represented with a table (CPT) or a Gaussian distribution for a discrete or continuous variable, respectively.

Conditional Independence

Figure 3.4 shows three BNs with different arrow directions over the same random variables a, b, and c.

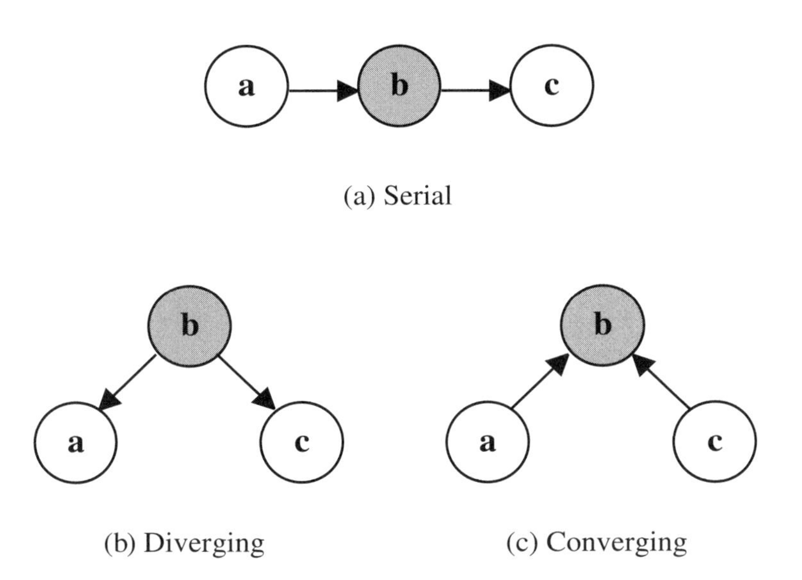

Fig. 3.4. Three BNs with different arrow directions over the same random variables a, b, and c. They appear in the case of serial, diverging, and converging connection, respectively.

Conditional independence appears in the case of serial and diverging connection, though one of the arrow directions has been reversed (see Figure 3.4(a) and (b)). Both networks have a common property: if the state of b is known, then no knowledge of c will alter the probability of a. In other words, the variables a and c are independent given the variable b (symbolized as "$a \parallel c \mid b$") if

$$P(a|b) = P(a|c, b). \tag{3.4}$$

By the conditioned Bayes's rule we get

$$\begin{aligned} P(a|c,b) &= \frac{P(c|a,b)P(a|b)}{P(c|b)} \\ &= \frac{P(c|b)P(a|b)}{P(c|b)} \\ &= P(a|b). \end{aligned} \tag{3.5}$$

The joint probability function, $P(a,b,c)$, of both networks can be written as

$$P(a,b,c) = P(a|c,b)P(c|b)P(b) = P(a|b)P(c|b)P(b). \tag{3.6}$$

This can be compared with the converging connection where a and c are parents of b (see Figure 3.4(c)). This network corresponds to the property "$a \parallel c$" (a and c are independent) but not to "$a \parallel c \mid b$" (a and c are independent given b); consequently,

$$P(a|c) = P(a), \tag{3.7}$$

and

$$P(a,c) = P(a)P(c), \tag{3.8}$$

so that the joint probability function, $P(a,b,c)$, of Figure 3.4(c) can be written as

$$P(a,b,c) = P(b|a,c)P(a,c) = P(b|a,c)P(a)P(c) \tag{3.9}$$

BN Joint PDF

In general, a BN joint PDF implicitly portrays factorizations that are simplifications of the chain rule of probability, namely

$$\begin{aligned} & P(Z_1, Z_2, ..., Z_K) \\ &= \prod_{k=1}^{K} P(Z_k|Z_{k+1}, ..., Z_K) \\ &= \prod_{k=1}^{K} P(Z_k|Pa(Z_k)). \end{aligned} \tag{3.10}$$

The first equality is derived from the probabilistic chain rule (see Eq. (3.3)) and the second equality holds under a particular BN, where $Pa(Z_k)$ denotes the parents of BN variable Z_k.

Figure 3.5 shows an example of a BN topology that describes the conditional relationship among a, b, c, d, e, f, g and h. By the probabilistic chain rule, we get

$$\begin{aligned}
&P(a, b, c, d, e, f, g, h) \\
&= P(a)P(b|a)P(c|a, b)P(d|a, b, c)P(e|a, b, c, d)P(f|a, b, c, d, e) \\
&\quad P(g|a, b, c, d, e, f)P(h|a, b, c, d, e, f, g)
\end{aligned} \tag{3.11}$$

and then by BN the factorization of the joint PDF can be simplified as

$$\begin{aligned}
&P(a, b, c, d, e, f, g, h) \\
&= P(a)P(b)P(c)P(d|a, b)P(e|b, c)P(f)P(g|e)P(h|e, f).
\end{aligned} \tag{3.12}$$

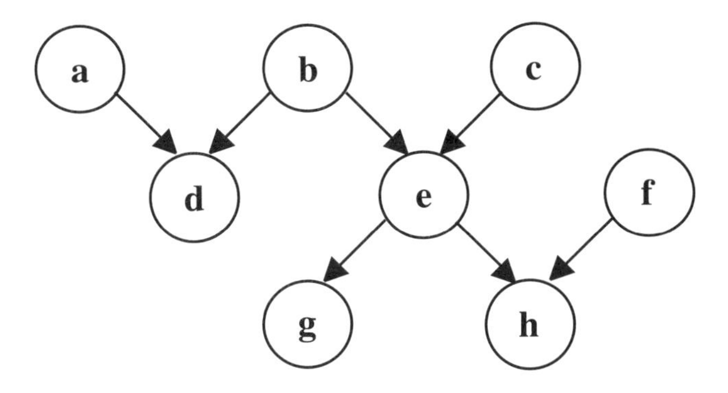

Fig. 3.5. Example of BN topology describing conditional relationship among a, b, c, d, e, f, g and h.

3.1.3 Junction Tree Algorithm

The are several algorithms that automatically perform an inference on the graph. Some of these directly operate on the directed graph. However, in many cases, performing inference by carrying out global computations directly on the graph has shown to be difficult (Roberts, 2005).

The most popular algorithm, known as the junction tree algorithm, uses graph theory to form undirected graphs that have the same information-coupling properties as the original graph but are simpler to deal with (Roberts, 2005). This approach allows a decomposition of the joint PDF into a linked set of local conditional PDFs. Accordingly, a simplified form of the model can be constructed and reliably estimated.

The key in junction tree algorithms (Roberts, 2005) is to form a graph that has the same global properties as the original graph and that allows local representation to avoid brute force inference. Forming the junction tree is known as compiling the Bayesian network, and this is achieved by applying several types of graphical transformation (Jensen, 1998; Huang and Darwiche, 1994), as follows:

1. Moralization of the graph.
2. Triangulation of the graph.
3. Identifying cliques in the graph.
4. Joining the cliques and forming the junction tree.

In the following sections, we deal with each of these graph representations in turn and consider the original graph outlined in Figure 3.5.

Moral Graph

The term "moralization" is coined from work in the United States (US), where having children out of wedlock is widely seen as "immoral" (Roberts, 2005). Thus, a moral graph is achieved by:

- Marrying the parents by adding a link between any pair of variables with a common child, and
- Forming an undirected graph by dropping the direction of the links.

From Figure 3.5, we can see that nodes [a] and [b] are parents of node [d], nodes [b] and [c] are parents of node [e], and that nodes [e] and [f] are parents of node [h]. Consequently, by adding a link between the node pairs [a,b], [b,c] and [e,f] and then dropping the direction of the links, the resulting graph (Figure 3.6) is called a **moral graph**.

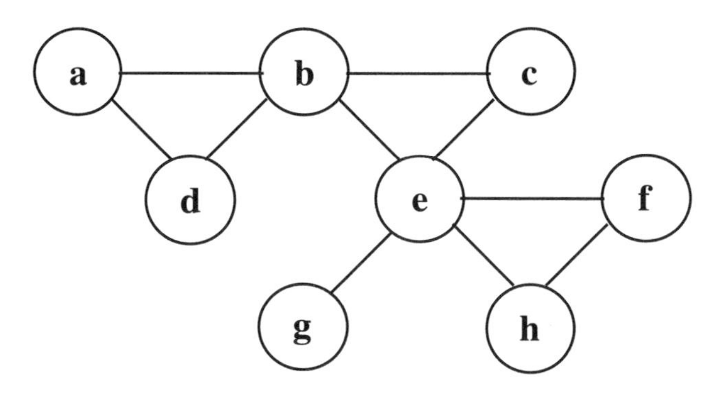

Fig. 3.6. Moral and triangulated graph of Figure 3.5.

Triangulated Graph

As described previously, the resulting junction tree should have the same global properties as the original graph, which in this case is the BN. Here, the algorithm introduces a triangulation step in which selected arcs are added to the moral graph so that sets of maximally connected subgraphs are formed. The idea is to obtain a representation based on a number of maximally connected subgraphs, which form the basis of local representation and which are connected such that the global properties are preserved (Roberts, 2005).

In practice, a graph is triangulated (chordal, decomposable) if there are no "chordless cycles", where a chord is an edge connecting two nonconsecutive vertices in a cycle of length > 3 (Bartels et al., 2005). This means that we need to form triplets of nodes. Since there is already no "chordless cycles" in Figure 3.6, we call the resulting graph a **moral and triangulated graph**.

Cliques

In graph theory, a maximally connected graph is also called a **clique**. To identify a clique, for each variable A with $Pa(A) \neq 0$ in the triangulated graph, we form a subset containing $Pa(A) \bigcup A$.

In our example of Figure 3.6, the cliques are formed as

$$\begin{aligned} C_1 &= [a, b, d] \\ C_2 &= [b, c, e] \\ C_3 &= [e, g] \\ C_4 &= [e, f, h] \end{aligned} \tag{3.13}$$

Junction Tree

In the last step, we join the cliques to form a **junction tree**. We start with cliques as the nodes, in which each link between two cliques is labeled by using a **separator** of a non-empty intersection between those cliques. The resulting graph (Figure 3.7) is called a **junction graph**, where the cliques are represented by the oval nodes and the separator sets are represented by the square nodes.

If the junction graph has cycles, then all separators on the cycles contain the same variables. Therefore, any of the links can be removed to break a cycle, and by removing the links we eventually obtain a tree, which is called a **junction tree** (see Figure 3.8).

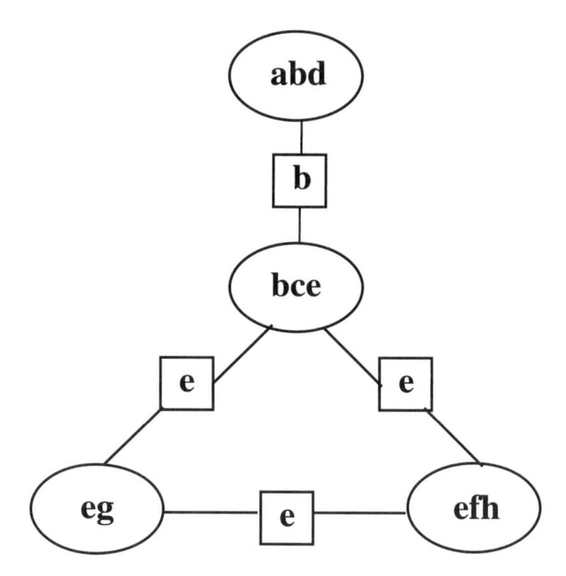

Fig. 3.7. Junction graph of Figure 3.5.

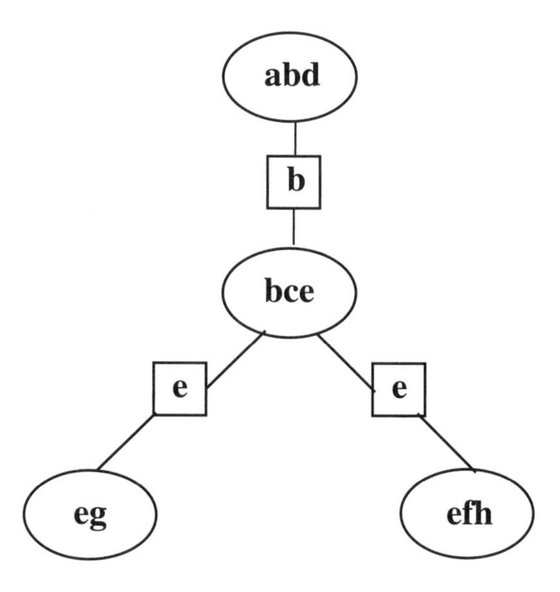

Fig. 3.8. The resulting junction tree.

Inference on Junction Tree

The joint probability function on a junction tree is defined as the product of all cluster potentials, divided by the product of the separator potentials (Huang and Darwiche, 1994), as follows:

$$P(Z_1, Z_2, ..., Z_K) = \frac{\prod_i \phi_{C_i}}{\prod_j \phi_{S_i}}, \tag{3.14}$$

where ϕ_{C_i} is the cluster potential (the probability over the cluster C_i), and ϕ_{S_i} is the separator potential (the probability over the separator S_i). Thus, from Figure 3.7 we get

$$\begin{aligned} P(a,b,c,d,e,f,g,h) &= \frac{\phi_{C_1}\phi_{C_2}\phi_{C_3}\phi_{C_4}}{\phi_{S_1}\phi_{S_2}\phi_{S_3}} \\ &= \frac{P(a,b,d)P(b,c,e)P(e,g)P(e,f,h)}{P(b)P(e)P(e)}. \end{aligned} \tag{3.15}$$

This clique potential is only a function of the local variables in the clique. It is obtained by considering just those nodes in the original BN. For example, considering clique $C_1 = [a, b, d]$, the three nodes in this clique yield the same structure as those shown in Figure 3.9.

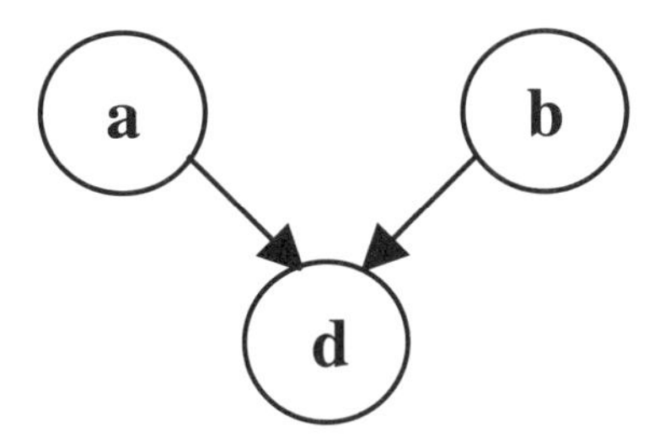

Fig. 3.9. Clique $C_1 = [a, b, d]$ in the original graph of Figure 3.5.

By applying the rules of directed graph models to this subgraph, we then obtain

$$\phi_{C_1} = P(a,b,d) = P(d|a,b)P(a)P(b). \tag{3.16}$$

Following the same procedure for cliques C_2, C_3 and C_4, we get

$$\begin{aligned}\phi_{C_2} &= P(b,c,e) = P(e|b,c)P(b)P(c),\\ \phi_{C_3} &= P(e,g) = P(g|e)P(e),\\ \phi_{C_4} &= P(e,f,h) = P(h|e,f)P(e)P(f).\end{aligned} \tag{3.17}$$

Thus the PDF over all the variables is

$$\begin{aligned}P(a,b,c,d,e,f,g,h) &= \frac{\phi_{C_1}\phi_{C_2}\phi_{C_3}\phi_{C_4}}{\phi_{S_1}\phi_{S_2}\phi_{S_3}}\\ &= \frac{P(a,b,d)P(b,c,e)P(e,g)P(e,f,h)}{P(b)P(e)P(e)}\\ &= \frac{[P(d|a,b)P(a)P(b)][P(e|b,c)P(b)P(c)]}{P(b)} \cdot\\ &\quad \frac{[P(g|e)P(e)][P(h|e,f)P(e)P(f)]}{P(e)P(e)}\\ &= P(a)P(b)P(c)P(d|a,b)P(e|b,c)P(f)P(g|e)P(h|e,f)\end{aligned} \tag{3.18}$$

which is exactly the global property we obtained in Eq. (3.12). This shows that the global properties are preserved even while local computation of the subgraphs proceeds.

3.2 Procedure of GFIKS

.

We now introduce the design of our proposed framework, GFIKS. It is based on the graphical model described in the previous section. In the statistics-based approach, given some observation data D, we train a model M. One key problem is the computing of the likelihood, $P(D|M)$, that predicts the data based on the current knowledge of the model.

We can model the probability density function, $P(D|M)$, in simple cases by using CPT (if D is discrete) or continuous functions such as Gaussian densities (if D is continuous); the output probability for given data d and model parameter m is then simply calculated as

$$p(d|m) = P(D = d|M = m). \tag{3.19}$$

Then, assume that we want to incorporate additional knowledge sources in the model. The procedure of GFIKS, as shown in the flow of Figure 3.10, consists of several steps:

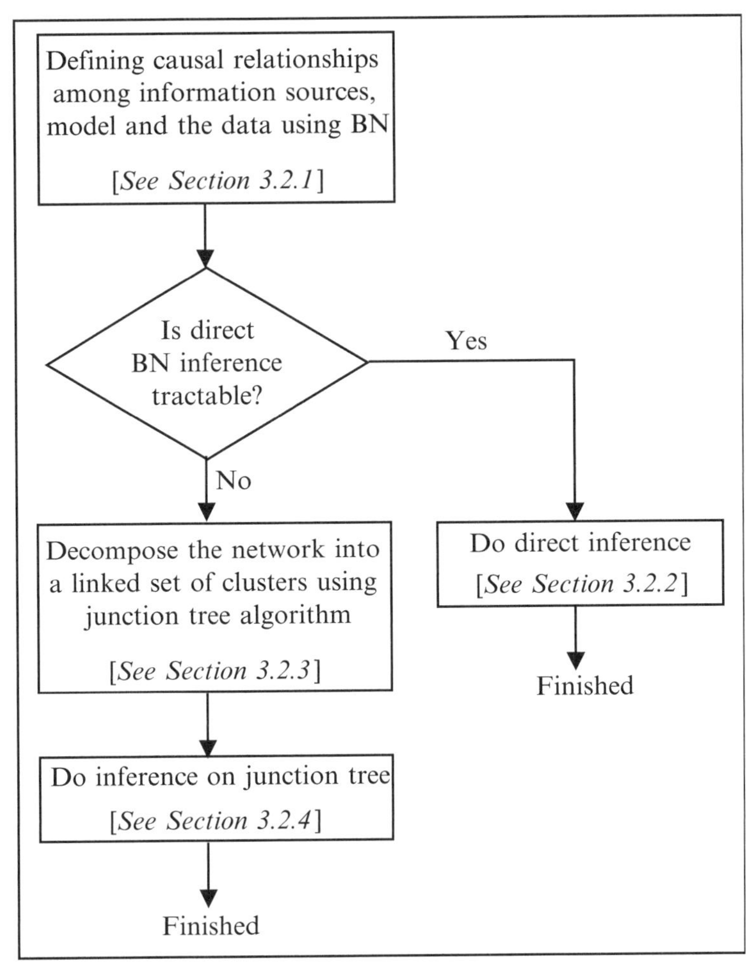

Fig. 3.10. General procedure of GFIKS (graphical framework to incorporate additional knowledge sources).

- Step 1:

  Designing causal relationships among information sources, the model, and the data using BN.

- Step 2:

  Analyzing the network. If direct BN inference is tractable, we can perform direct inference on BN. Otherwise, go to Step 3.

- Step 3:

  If direct BN inference is intractable, the network has to be decomposed into a linked set of clusters using the junction tree algorithm described in Section 3.1.3.

- Step 4:

  Carrying out inference on the junction tree.

Each step is described in more detail in the following sections.

3.2.1 Causal Relationship between Information Sources

Let us start from a simple case, where the causal relationship between D and M is described using BN, like the one shown in Figure 3.11(a); here, we assume M to be a discrete variable denoted by the square node and D to be a continuous variable denoted by the oval node.

The BN joint probability function can be factorized as Eq. (3.10), and thus we obtain

$$P(D, M) = P(D|M)P(M), \tag{3.20}$$

from Figure 3.11(a). Accordingly, we simply define the conditional relationship among D, M, and K to incorporate additional knowledge K in $P(D, M)$, and then we express the joint probability model in a similar way. The conditional relationship among D, M, and K, for example, can be described by the BN in Figure 3.11(b). Consequently, the BN joint probability function becomes

$$P(D, K, M) = P(D|K, M)P(K|M)P(M). \tag{3.21}$$

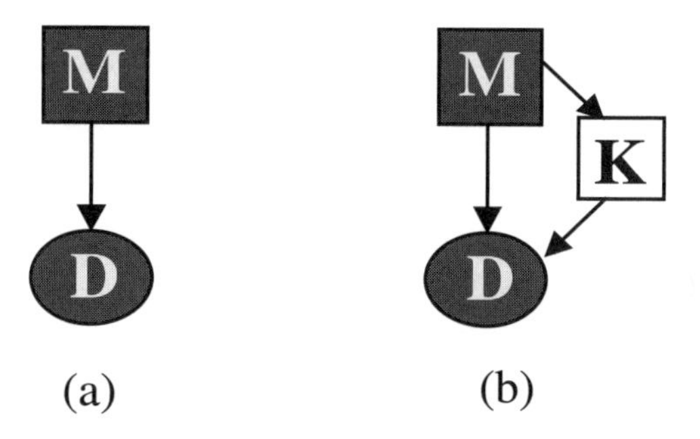

Fig. 3.11. (a) BN topology describing the conditional relationship between data D and model M. (b) BN topology describing the conditional relationship among D, M, and additional knowledge K.

Now, let us consider a more detailed example, where we assume that there are $K_1, K_2, ..., K_N$ knowledge sources. Here, we also assume that they are all conditionally independent. Figure 3.12 shows two examples of conditional relationship structures for D, M, and $K_1, K_2, ..., K_N$.

Then, the joint PDF becomes

$$\begin{aligned} &P(D, K_1, ..., K_N, M) \\ &= P(D|K_1, ..., K_N, M)P(K_1|M)...P(K_N|M)P(M) \end{aligned} \tag{3.22}$$

for the BN of Figure 3.12(a), according to Eq. (3.10). If there are some K_i that receive no causal impact from M, as shown in Figure 3.12(b) (see K_1 and K_N), then the joint probability function becomes

$$\begin{aligned} &P(D, K_1, ..., K_N, M) \\ &= P(D|K_1, ..., K_N, M)P(K_1)P(K_2|M)...P(K_N)P(M). \end{aligned} \tag{3.23}$$

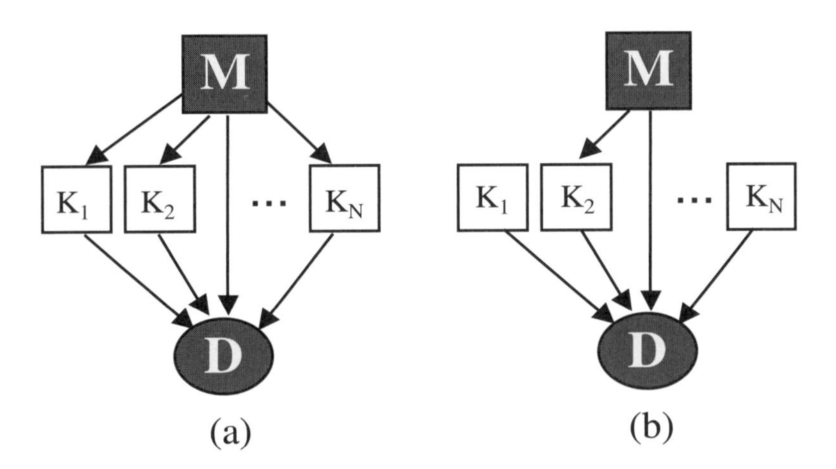

Fig. 3.12. Examples of BN topologies describing the conditional relationship among data D, model M, and several knowledge sources $K_1, K_2, ..., K_N$.

As can be seen, different conditional independence assumptions may lead to different probability function decompositions (see Eqs. (3.22) and (3.23)).

3.2.2 Direct Inference on Bayesian Network

Our primary interest during inference is to calculate the global conditional probability, $P(D|K_1, ..., K_N, M)$. If direct calculation of the PDF is possible to perform, the following two cases may be considered.

1. **All variables can be observed.**

   In this case, the PDF may simply be calculated as in Eq. (3.19)

$$p(d|k_{1_j},...,k_{N_j},m) \\ = P(D=d|K_1=k_{1_j},...,K_N=k_{N_j},M=m). \tag{3.24}$$

2. **Some variables, such as $K_1,...,K_N$ of additional knowledge sources cannot be observed or are hidden.**

   In this case, the PDF calculation is done using Eq. (3.22) and by marginalization over all possible $K_i : k_{i_1}, k_{i_2}, ..., k_{i_M}$ for all K_i

$$\begin{aligned} p(d|m) &= \frac{p(d,m)}{p(m)} \\ &= \frac{\sum_{1_j=1}^{M_1} \cdots \sum_{N_j=1}^{M_N} p(d,k_{1_j},...,k_{N_j},m)}{p(m)} \\ &= \sum_{1_j=1}^{M_1} \cdots \sum_{N_j=1}^{M_N} p(d|k_{1_j},...,k_{N_j},m)p(k_{1_j}|m)...p(k_{N_j}|m), \end{aligned} \tag{3.25}$$

   where for simplicity, we use d, m and k_{i_j} instead of $\langle D=d \rangle$, $\langle M=m \rangle$, and $\langle K_i = k_{i_j} \rangle$.

However, the calculation of the global conditional probability $P(D|K_1,...,K_N,M)$ is occasionally not trivial, due to the significant number of variables and/or the computational complexity. In this case, directed graphs need to be decomposed into clusters of variables, on which the relevant computations can be carried out. This may be done with the junction tree algorithm (Jensen, 1998), which is briefly described in the following section.

3.2.3 Junction Tree Decomposition

Consider a simple case where we only incorporate two additional knowledge sources, K_1 and K_2. The causal relationship among D, M, K_1, and K_2 is described by the BN in Figure 3.13(a). Here, M, K_1, and K_2 are discrete variables denoted by the square nodes, and D is a continuous variable denoted by the oval node.

According to the junction tree algorithm described in Section 3.1.3, several graphical transformations have to be applied in order to obtain a junction tree. Figure 3.13(b) shows a moral and triangulated version of the BN from Figure 3.13(a). However, we can only obtain one cluster/clique with the full set of variables $\{D, M, K_1, \text{and } K_2\}$ from this triangulated graph without the ability to decompose this set any further. Fortunately, since K_1 and K_2 are assumed to be independent, we can obtain an equivalent graph, like the one in Figure 3.13(c), by reversing some arrows. Figure 3.13(d) shows the moral and triangulated version of this graph. We can then identify the clusters/cliques and obtain the junction tree outlined in Figure 3.13(e), where the cluster sets are represented by the oval nodes and the separator sets are represented by the square nodes.

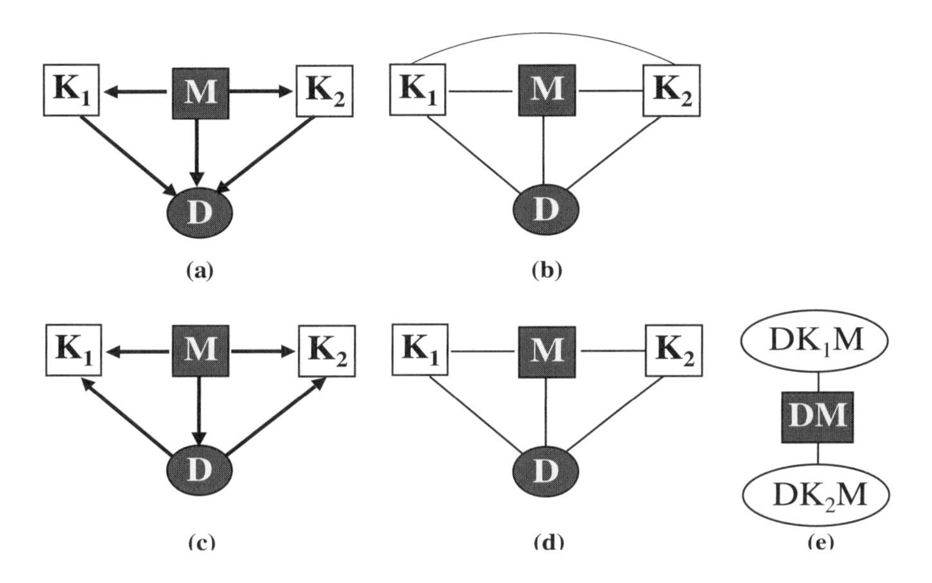

Fig. 3.13. (a) BN topology describing conditional relationship among D, M, K_1, and K_2. (b) Moral and triangulated graph of Figure 3.13(a). (c) Equivalent BN topology. (d) Moral and triangulated graph of Figure 3.13(c). (e) Junction tree of Figure 3.13(d).

Thus, the joint probability function, $P(D, K_1, K_2, M)$, according to Eq. (3.14), becomes

$$P(D, K_1, K_2, M) = \frac{P(D, K_1, M)P(D, K_2, M)}{P(D, M)}, \tag{3.26}$$

according to Figure 3.13(e), where $P(D, K_1, M)$ and $P(D, K_2, M)$ are the cluster potentials and $P(D, M)$ is the separator potential.

The equivalent BN topology of the BN shown in Figure 3.12(a) can be described as in Figure 3.14(a) based on similar assumptions and considerations. The corresponding junction tree is given in Figure 3.14(b), where there are N clusters of variables $\{\{D, K_1, M\}, \{D, K_2, M\}, ...\{D, K_N, M\}\}$ and $N-1$ separators $\{D, M\}$.

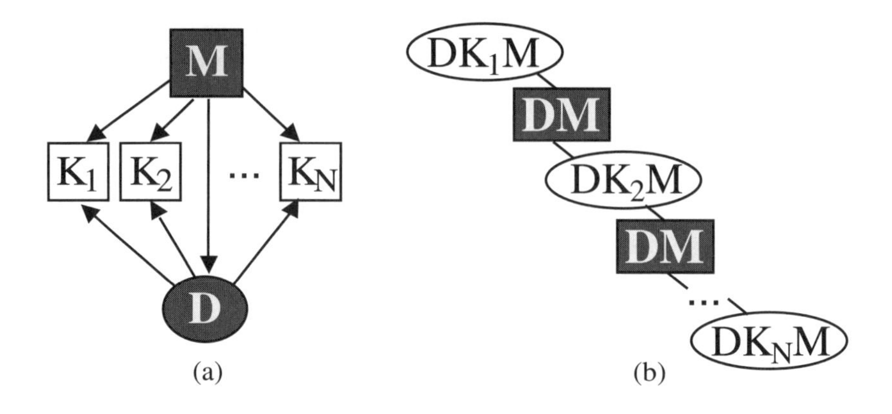

Fig. 3.14. (a) Equivalent BN topology of the BN shown in Figure 3.12(a). (b) Corresponding junction tree.

The joint probability function of Eq. (3.22) may then be decomposed into

$$\begin{aligned} P(D, K_1, ..., K_N, M) \\ &= \frac{\prod_{i=1}^{N} P(D, K_i, M)}{\prod_{i=1}^{N-1} P(D, M)} \\ &= \frac{\prod_{i=1}^{N} P(D, K_i, M)}{P(D, M)^{N-1}}. \end{aligned} \tag{3.27}$$

This provides a new way of representing, $P(D, K_1, ..., K_N, M)$, the joint probability function as a composition of several local joint probability functions $P(D, K_1, M), ..., P(D, K_N, M)$, which correspond to the probability of observational data D given the specific additional knowledge of $K_1, K_2, ..., K_N$.

3.2.4 Junction Tree Inference

We obtain

$$P(D, K_i, M) = P(D|K_i, M)P(K_i|M)P(M), \tag{3.28}$$

for all $P(D, K_i, M)$ using the chain rule, and thus Eq. (3.27) becomes

$$\begin{aligned} & P(D, K_1, ..., K_N, M) \\ & = \frac{\prod_{i=1}^{N} P(D, K_i, M)}{P(D, M)^{N-1}} \\ & = \frac{\prod_{i=1}^{N} \{P(D|K_i, M)P(K_i|M)P(M)\}}{\{P(D|M)P(M)\}^{N-1}} \\ & = \frac{\prod_{i=1}^{N} P(D|K_i, M)}{P(D|M)^{N-1}} P(K_1|M)...P(K_N|M)P(M). \end{aligned} \tag{3.29}$$

Comparing this with Eq. (3.22), we observe that

$$P(D|K_1, ..., K_N, M) = \frac{\prod_{i=1}^{N} P(D|K_i, M)}{P(D|M)^{N-1}}, \tag{3.30}$$

which indicates that $P(D|K_1, ..., K_N, M)$ may be decomposed into separate terms, corresponding to the probability of observing data D given the specific additional knowledge of $K_1, K_2, ..., K_N$.

It is now much easier to define, estimate, and calculate several simple instances of $P(D|K_i, M)$ rather than a single but complex $P(D|K_1, ..., K_N, M)$.

The output probability during inference for given data d, model parameter m, and additional knowledge source k_{1_j} is then calculated as

$$p(d|k_{1_j}, ..., k_{N_j}, m) = \frac{\prod_{i=1}^{N} P(D = d|K_i = k_{i_j}, M = m)}{P(D = d|M = m)^{N-1}}. \tag{3.31}$$

3.3 Practical Issues of GFIKS

3.3.1 Types of Knowledge Sources

As described in Section 1.4, there exist numerous sources of variability contained in the speech signal that have to be handled by an ASR, in order to successfully parse a speech stream and extract sequences of words. Thus, to examine the efficiency of our proposed framework in handling this issue, we intend to incorporate various knowledge sources that come from different domain, including:

1. Contextual variability:

   Wide phonetic-context information, i.e., previous and following context information.

2. Speaker variability:

   - Gender information,
     i.e., female and male.
   - Accent information,
     i.e., American English and Australian English.

3. Environmental variability:

   - noise type information,
     i.e., subway, babble, car, exhibition hall, restaurant, street, airport and train station noise.
   - noise level information,
     i.e., signal-to-noise ratio (SNR) from clean to -5dB.

By applying GFIKS, we attempt to incorporate, not only a single type of knowledge source, but also the possibility to combine different type of knowledge sources.

3.3.2 Different Levels of Incorporation

As mentioned in Section 2.3.5, statistical speech recognition consists of multi-level probability estimation, including estimations at the state, phonetic-unit, word and sentence levels. Here, we apply our framework starting from the lowest level of the ASR system, i.e., the acoustic model, because it is arguably the central part of any speech recognition system (Huang et al., 2001).

Therefore, we attempt to incorporate various knowledge sources, at several levels of the acoustic model, including:

1. HMM states
2. HMM phonetic unit models

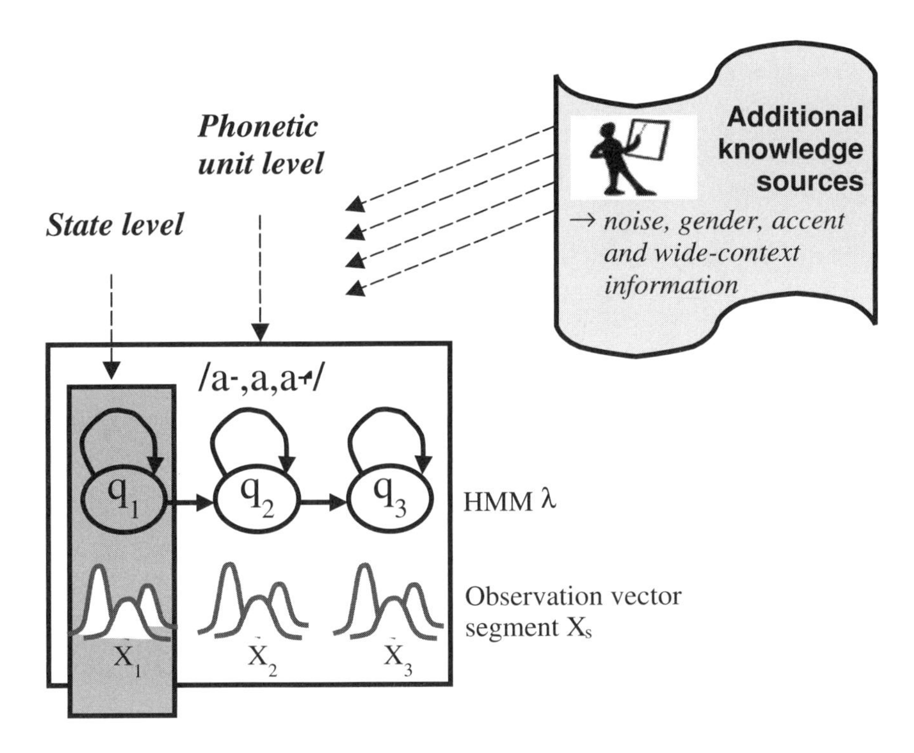

Fig. 3.15. Incorporating knowledge sources into HMM state (denoted by a small box) and phonetic unit level (denoted by a large box).

The scheme of incorporation is illustrated in Figure 3.15. We first focus on how the additional sources of knowledge are incorporated in the HMM state distribution, denoted by the small box in the figure, and then we focus on how the additional sources of knowledge are incorporated in HMM phonetic modeling, denoted by the large box in the figure. Both issues are described in the following chapter.

4

Speech Recognition Using GFIKS

In this chapter, we demonstrate how the statistical speech recognition system may incorporate additional sources by utilizing GFIKS at different levels, HMM state and phonetic-unit. We also present some experimental results of incorporating various knowledge sources, including environmental variability (i.e., background noise information), speaker variability (i.e., accent and gender information) and contextual variability (i.e., wide-phonetic information). The incorporation of these knowledge sources may be done only for a single type of knowledge source, or even the combination between different type of knowledge sources.

We describe some common considerations of using GFIKS at the HMM state level in Section 4.1 and at the HMM phonetic-unit level in Section 4.2. These issues include defining causal relationships between information sources, inference, training issues, and enhancing model reliability. Then, in Section 4.3, we describe an experimental evaluation of applying the proposed GFIKS to the task of incorporating various knowledge sources. Finally, in Section 4.4, the summary of the experiments are presented and the comparison between different level of incorporation is also discussed.

4.1 Applying GFIKS at the HMM State Level

To apply GFIKS at the HMM state level, we focus solely only on a single HMM state, as shown by the shaded box in Figure 4.1(a). Corresponding to the illustration in Figure 3.11(a), the BN topology structure is shown in Figure 4.1(b), where model M is now our HMM state Q, and D is observation variable X. Following our presentation of the GFIKS procedure (Section 3.2), we define the causal relationship and inference.

S. Sakti et al., *Incorporating Knowledge Sources into Statistical Speech Recognition*,
Lecture Notes in Electrical Engineering 42, DOI: 10.1007/978-0-387-85830-2_4,

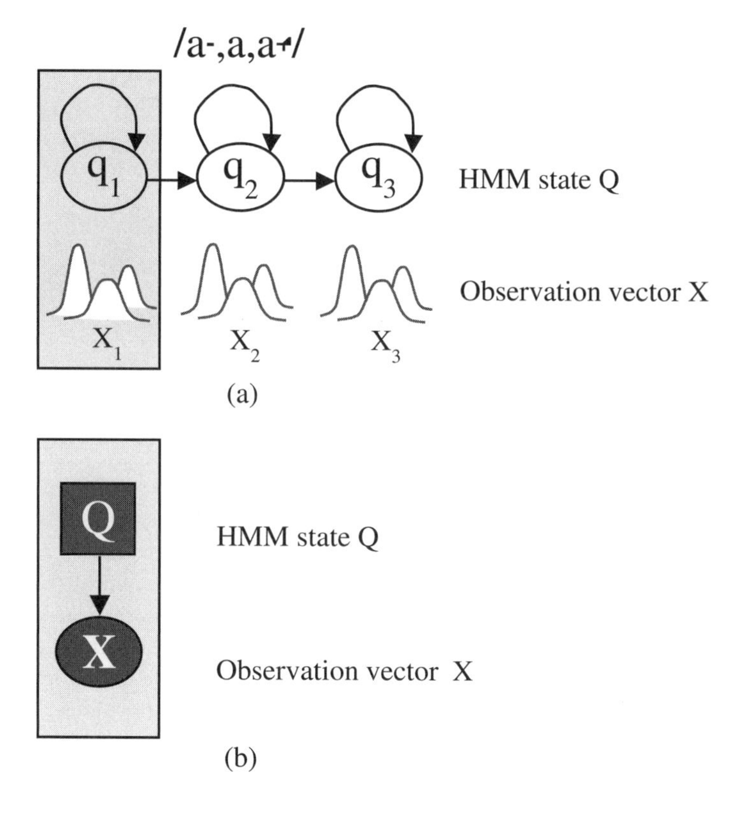

Fig. 4.1. (a) Applying GFIKS at the HMM state level. (b) BN topology structure describing the conditional relationship between HMM state Q and observation vector X.

4.1.1 Causal Relationship Between Information Sources

Given the BN topology structure shown in Figure 4.1(b), the HMM state PDF is now represented by the BN joint probability function, which is similar to Eq. (3.20):

$$P(X, Q) = P(X|Q)P(Q). \tag{4.1}$$

We can simply follow Eq. (3.22), giving

$$\begin{aligned} &P(X, K_1, ..., K_N, Q) \\ &= P(X|K_1, ..., K_N, Q)P(K_1|Q)...P(K_N|Q)P(Q), \end{aligned} \tag{4.2}$$

to incorporate additional knowledge sources $K_1, K_2, ..., K_N$ in our HMM state distribution $P(X, Q)$ (assuming that all $K_1, K_2, ..., K_N$ are independent given Q) as shown in Figure 4.2.

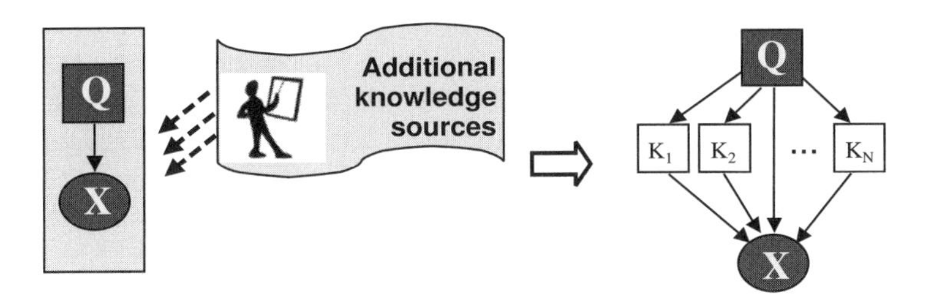

Fig. 4.2. BN topology structure after incorporating additional knowledge sources $K_1, K_2, ..., K_N$ in HMM state distribution $P(X, Q)$ (assuming that all $K_1, K_2, ..., K_N$ are independent given Q).

4.1.2 Inference

Our primary interest is to calculate the $P(X|K_1, ..., K_N, Q)$ of the HMM state output probability, that can easily be modeled with a Gaussian function. We thus directly obtain the state output.

If all additional knowledge sources $K_1, ..., K_N$ are assumed to be hidden as described in Section 3.2.2, the state output probability is obtained as indicated in Eq. (3.25), by marginalization over all possible $K_i : k_{i_1}, k_{i_2}, ..., k_{i_M}$ for all K_i, $1 \leq i \leq N$:

$$p(x_t|q_t) = \sum_{1_j=1}^{M_1} ... \sum_{N_j=1}^{M_N} p(x_t|k_{1_j}, ..., k_{N_j}, q_t)p(k_{1_j}|q_t)...p(k_{N_j}|q_t). \quad (4.3)$$

We observe that Eq. (4.3) is equivalent to the state output probability of the conventional HMM of Eq. (2.43) if we treat term $p(k_{1_j}|q_t)...p(k_{N_j}|q_t)$ as a mixture weight coefficients for the Gaussian component $P(X|k_{1_j}, ..., k_{N_j}, q_t)$. Since expressions such as Eq. (4.3) represent a mixture of Gaussians, we are able to perform recognition using existing HMM-based decoders without the need for any modification. Furthermore, since the BN is used only to infer the state output likelihood, we can retain our HMM-based acoustic model topology, where HMM state transitions are still used to govern temporal speech characteristics. This approach is also known as the hybrid HMM/BN modeling framework and is described in (Markov and Nakamura, 2006; Markov et al., 2003; Markov and Nakamura, 2005). We call this the model obtained by incorporating additional knowledge at the state level the HMM/BN model.

4.1.3 Enhancing Model Reliability

According to Eq. (4.3), for each value k_{i_j} of the additional knowledge source K_i, there is a corresponding Gaussian mixture component. An example of observation space modeling by BN with one additional knowledge source K_i

is shown in Figure 4.3. If all instances of K_i yield M values, then the number of Gaussians for each K_i may be M. Therefore, the total number of Gaussians for each state with N knowledge may become M^N. If the amount of training data is not sufficient to obtain a reliable estimate of the increased model parameters, the overall performance may significantly degrade. It is thus necessary to reduce the number of Gaussians. Any type of clustering technique, e.g., knowledge-based or data-driven clustering, can be applied here.

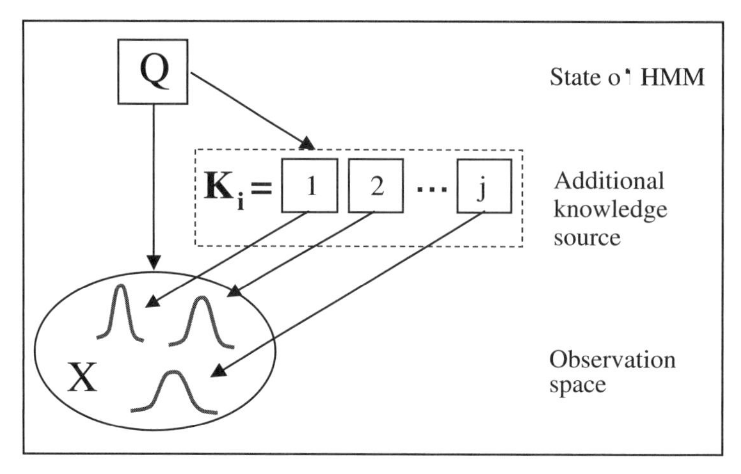

Fig. 4.3. Example of observation space modeling by BN, where each value of K_i corresponds to a different Gaussian.

4.1.4 Training and Recognition Issues

The parameter training of the HMM/BN model is based on the forward-backward algorithm, which consists of BN training and updating of the HMM transition probabilities. This algorithm consists of the following steps:

1. Initialization:
   Initialize HMM/BN parameters using the bootstrapping of the conventional HMM model.

2. Forward-Backward algorithm:
   Obtain a time-aligned state segmentation of the training data.

3. BN training:
   Train the BN using state-labeled training data.

4. Transition probability updating.

5. Embedded BN/HMM training.

6. Convergence check:
   Stop if the convergence criterion is met, otherwise go to step 2.

The training of the state BN in step 3 is done using standard statistical methods. ML parameter estimation is applied when all variables are observable during training; however, if some variables are hidden, the parameters may then be estimated using the standard EM algorithm.

For the recognition issues, all of the additional knowledge sources $K_1, ..., K_N$ are assumed to be hidden, so the state output probability can be treated as the state output probability of the conventional HMM (see Section 4.1.1). Consequently, we simply undertake recognition using existing HMM based decoders without the need for any modifications.

4.2 Applying GFIKS at the HMM Phonetic-unit Level

To apply GFIKS at the HMM phonetic-unit level, we focus on a single phonetic unit from the entire HMM as shown in the shaded box in Figure 4.4(a). Similar to the configuration shown in Figure 3.11(a), the BN topology structure of the statistical acoustic model is described in Figure 4.4(b), where M is currently our HMM phonetic model λ, and D is observation segment X_s. Then, following the GFIKS procedure given in Section 3.2, we define the causal relationship and inference in the following sections.

4.2.1 Causal Relationship between Information Sources

According to the BN topology structure shown in Figure 4.4(b), the probability function of HMM phonetic units is now represented by the BN joint probability function, similar to Eq. (3.20):

$$P(X_s, \lambda) = P(X_s|\lambda)P(\lambda). \tag{4.4}$$

We can simply follow Eq. (3.22), giving

$$\begin{aligned} &P(X_s, K_1, ..., K_N, \lambda) \\ &= P(X_s|K_1, ..., K_N, \lambda)P(K_1|\lambda)...P(K_N|\lambda)P(\lambda), \end{aligned} \tag{4.5}$$

to incorporate additional knowledge sources $K_1, K_2, ..., K_N$ in our HMM phonetic model, $P(X_s, \lambda)$ (assuming that all $K_1, K_2, ..., K_N$ are independent given λ), as shown in Figure 4.5.

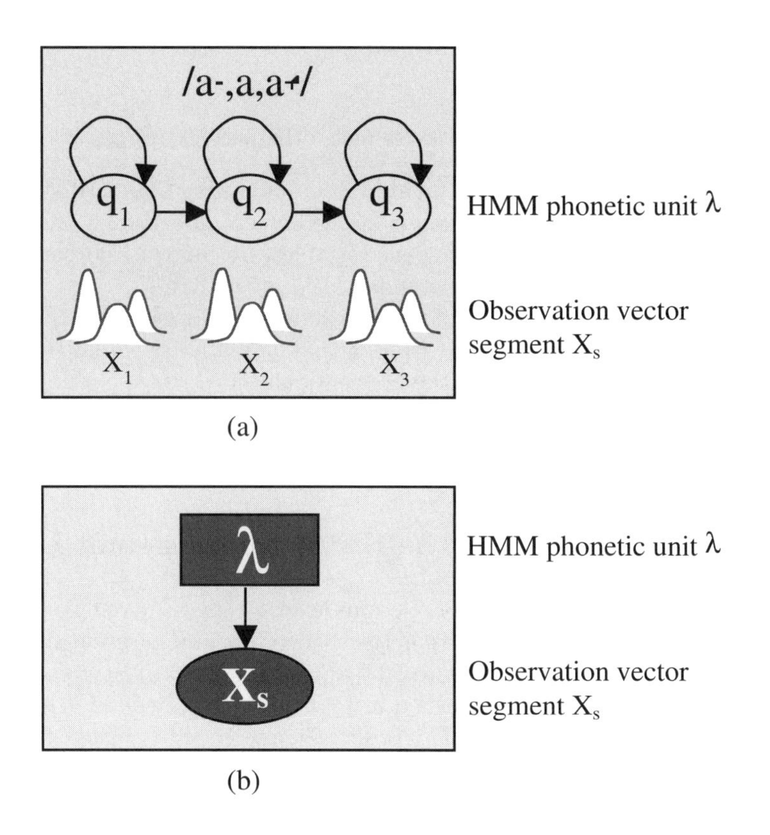

Fig. 4.4. (a) Applying GFIKS at the HMM phonetic-unit level. (b) BN topology structure describing the conditional relationship between HMM phonetic model λ and observation segment X_s.

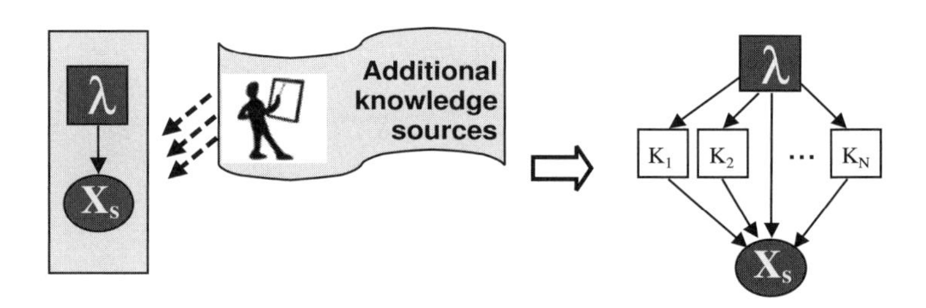

Fig. 4.5. BN topology structure after incorporating additional knowledge sources $K_1, K_2, ..., K_N$ in HMM phonetic model $P(X_s, \lambda)$ (assuming that all $K_1, K_2, ..., K_N$ are independent given λ).

4.2.2 Inference

Our primary interest is to calculate $P(X_s|K_1, ..., K_N, \lambda)$ given input segment X_s. However, it is difficult to obtain a simple functional form for this conditional PDF, because it involves an HMM model, λ, and a segment, X_s, of variable duration. Accordingly, we need to decompose $P(X_s|K_1, ..., K_N, \lambda)$ by the junction tree algorithm as described in Section 3.1.3. It can be decomposed as

$$P(X_s|K_1, K_2, ..., K_N, \lambda) = \frac{\prod_{i=1}^{N} P(X_s|K_i, \lambda)}{P(X_s|\lambda)^{N-1}}, \tag{4.6}$$

according to Eq. (3.30), which indicates a new way of representing the HMM phonetic likelihood $P(X_s|K_1, K_2, ..., K_N, \lambda)$ through the composition of several less complex dependencies, i.e., $P(X_s|K_1, \lambda), ..., P(X_s|K_N, \lambda)$. This corresponds to the likelihood of segment observation data X_s, given the respective specific additional knowledge of K_1,K_2,..., or K_N.

4.2.3 Enhancing the Model Reliability

The estimates of parameters of knowledge-rich model $P(X_s|K_1, K_2, ..., K_N, \lambda)$ (even for the composition model) may become unreliable for insufficient amounts of training data; this may also be the case for the state output. The common approach to improve model reliability is to apply a smoothing technique, such as back-off or interpolation. We have investigated three different approaches:

1. "No decision":
   In this case, no smoothing technique is applied. We always accept the output value from the composition model $P(X_s|K_1, K_2, ..., K_N, \lambda)$ as the final output, so that
   $$P(X_s|\lambda) = P(X_s|K_1, K_2, ..., K_N, \lambda). \tag{4.7}$$
2. "Hard decision":
   Here, we accept only the output value from $P(X_s|K_1, K_2, ..., K_N, \lambda)$ of the composition model when it is larger than the output from the base model $P(X_s|\lambda)$. Otherwise, we fall back to $P(X_s|\lambda)$. This is similar to the back-off technique, but in this case, the back-off weight is either 0 or 1.
   $$P(X_s|\lambda) = \begin{cases} P(X_s|K_1, K_2, ..., K_N, \lambda), \\ \qquad \text{if } P(X_s|K_1, K_2, ..., K_N, \lambda) \geq P(X_s|\lambda) \\ P(X_s|\lambda), \\ \qquad \text{otherwise} \end{cases} \tag{4.8}$$
3. "Soft decision":
   Here, we use deleted interpolation, which is described below.

4.2.4 Deleted Interpolation

Deleted interpolation (DI) is an efficient technique that allows us to fall back to the more reliable model when the supposedly more precise model is, in fact, unreliable (Huang et al., 2001).

The concept involves interpolating two separately trained models with one being more reliably trained than the other. However, instead of interpolating the two models, we apply this approach to interpolating two phonetic likelihoods, where the phonetic likelihood of the composition model, $P(X_s|K_1, K_2, ..., K_N, \lambda)$, is the precise one, while the conventional HMM likelihood, $P(X_s|\lambda)$, is the more reliable one; accordingly, the interpolation phonetic likelihood, $P(X_s|\lambda)$, is obtained as

$$P(X_s|\hat{\lambda}) = \alpha P(X_s|K_1, K_2, ..., K_N, \lambda) + (1-\alpha)P(X_s|\lambda), \quad (4.9)$$

where α represents the weight of the HMM phonetic likelihood of the proposed composition model, and $(1-\alpha)$ represents the weight of the HMM phonetic likelihood of the conventional HMM model. If there is a sufficiently large amount of training data, $P(X_s|K_1, K_2, ..., K_N, \lambda)$ becomes more reliable and α is expected to tend towards 1.0. Otherwise, α will tend towards 0.0 so as to fall back to the more reliable model, $P(X_s|\lambda)$.

The optimal value of interpolation weights may be estimated using a development rather than a training set or using the cross-validation method (Huang et al., 2001). In this method, the training data is divided into M parts, and models are trained from each combination of $M-1$ parts, with the deleted part serving as development data to estimate the interpolation weights. These M sets of interpolation weights are then averaged to obtain the final weight.

4.2.5 Training and Recognition Issues

All components of the composition model $P(X_s|K_1, \lambda), ..., P(X_s|K_N, \lambda)$ have been trained separately given the segment observation data X_s and the respective specific additional knowledge of K_1,K_2,..., or K_N.

The implementation of the proposed composition model in an ASR system requires a special decoder that can work with several models. This can be avoided if the proposed models are applied by rescoring the N-best list generated by a standard HMM-based decoding system. Figure 4.6 shows a block diagram of such a rescoring procedure.

For each utterance in the test data, an N-best recognition (at the word level) of the baseline system is performed using a conventional HMM model and standard two-pass Viterbi decoding. Each N-best hypothesis includes an AM score, a LM score and a Viterbi segmentation of each phoneme. Then each phoneme segment in each hypothesis is rescored using the composition model. In each rescoring, we applied "no decision," "hard decision," and "soft decision" mechanisms to enhance the model reliability (see Section 4.2.3). Then, the new scores are combined with the LM score for this hypothesis. The hypothesis achieving the highest total utterance score among the N-best is selected as the new recognition output.

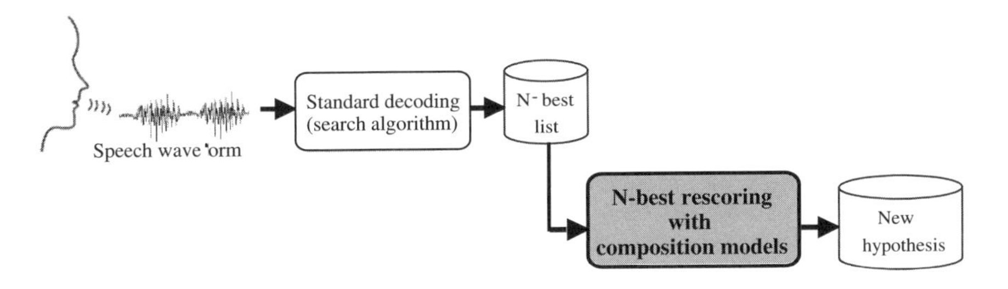

Fig. 4.6. Rescoring procedure with the composition models.

4.3 Experiments with Various Knowledge Sources

4.3.1 Incorporating Knowledge at the HMM State Level

We have incorporated various additional knowledge sources, including background noise, accent, gender and wide-phonetic knowledge information, by following a common method of applying GFIKS at the HMM state level as described in Section 4.1.

Incorporating Gender Information

Every individual person speaks differently. His/her speech reflects physical characteristics including age, gender, dialect, health, education and personal style (Huang et al., 2001). This is most noticeably expressed in the difference between the pitch levels produced by men and women. In general, women speak with a relatively high-pitched voice and men with a low-pitched voice. By applying GFIKS as described in the procedure below, we can incorporate gender information at each HMM state.

a. **Causal Relationship and Inference**

When we incorporate gender information G in a conventional HMM state, the BN topology for each HMM state may come to resemble the one shown in Figure 4.7, where X has the two parents Q and G.

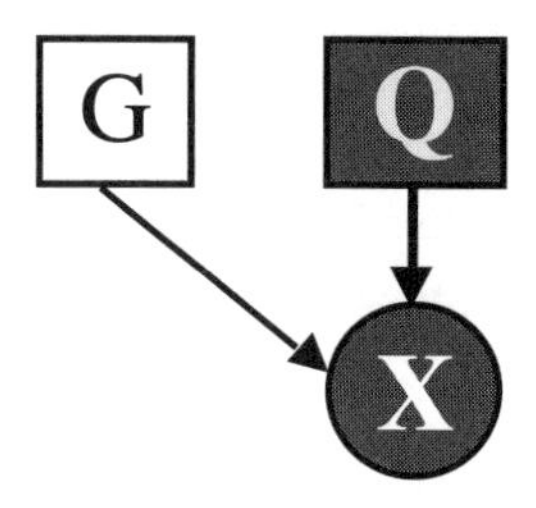

Fig. 4.7. BN topology structure showing the conditional relationship among HMM state Q, observation vector X, and additional knowledge source of gender information G.

The HMM state PDF is now the BN joint probability model, which is expressed as:

$$P(X, G, Q) = P(X|G, Q)P(G)P(Q), \tag{4.10}$$

where the PDF also depends on gender information G. When G is observable during inference, the HMM state output probability is simply

$$p(x_t|g_m, q_j) = P(X = x_t|G = g_m, Q = q_j). \tag{4.11}$$

However, when gender information G is assumed hidden (unknown) during recognition, the state output probability is calculated by marginalization over G:

$$p(x_t|q_j) = \sum_{m=1}^{M} p(g_m)p(x_t|g_m, q_j), \tag{4.12}$$

where, for simplicity, we use the x_t, q_j, and g_m notations instead of $\langle X = x_t \rangle$, $\langle Q = q_j \rangle$, and $\langle G = g_m \rangle$, respectively. $p(g_m)$ is the probability that the state q_j has gender g_m (male or female), and $p(x_t|g_m, q_j)$ is the probability of observation x_t given that we are in state q_j having gender g_m. Here, we can see that Eq. (4.12) is equivalent to the state output probability of the conventional HMM of Eq. (2.43), provided we treat the term $p(g_m|q_j)$ as a mixture weight coefficient for the Gaussian component $P(X|g_m, q_j)$.

b. **Enhancing Model Reliability**

As described in Section 4.1.3, for each value k_{i_j} of the additional knowledge source K_i, there is a corresponding Gaussian mixture component.

Here, gender G has two values: M for male and F for female. This means that the model will only have two gaussian mixture components per state. Thus, no type of clustering technique needs to be performed here.

c. **Training and Recognition Issues**

The training procedure is based on the algorithm described in Section 4.1.4. Since all variables, that is, HMM state Q, gender information G and feature variable X, are observable during training, only simple ML parameter estimation is applied to the training of the state BN in step 3 of the algorithm.

Recognition is performed using the existing HMM-based decoders without any modification by applying the above expression (Eq. (4.12)), since it represent a mixture of Gaussians as used in a standard triphone HMM acoustic model.

d. **Experimental Set-Up**

The experiments were conducted on the Wall Street Journal (WSJ0 and WSJ1) speech corpus task (Paul and Baker, 1992), closely following the evaluation scenario suggested by "The 1993 Hub and Spoke Paradigm for Continuous Speech Recognition Evaluation" (Kubala et al., 1994; Pallett et al., 1994). A brief explanation of the WSJ corpus task may also be found in Appendix A.3.

The pronunciation dictionary and LM used here are those used in the official evaluation. However, since the feature parameter and phoneme-unit set of an HMM model are not strictly defined in the official evaluation, we used the basic ATR parameter set-up. These parameters are defined as follows:

- Front-End Parameters

  For feature extraction, we used a sampling frequency of 16 kHz, a frame length of a 20-ms Hamming window, a frame shift of 10 ms, and 25 dimensional feature parameters consisting of 12-order MFCC, Δ MFCC and Δ log power.

- Unit Set

  The phoneme set used here consists of 43 English phonemes plus one silence (SIL). These are listed in Table 4.1.

Table 4.1. *English phoneme set.*

Phn	E.g.	Trans	Phn	E.g.	Trans
AA	*odd*	AA D	JH	*gee*	JH IY
AE	*at*	AE T	K	*key*	K IY
AH	*hut*	HH AH T	L	*lee*	L IY
AO	*ought*	AO T	M	*me*	M IY
AW	*cow*	K AW	N	*knee*	N IY
AX	*of*	AX V	NG	*ping*	P IH NG
AXR	*are*	AXR	OW	*oat*	OW T
AY	*hide*	HH AY D	OY	*toy*	T OY
B	*be*	B IY	P	*pee*	P IY
CH	*cheese*	CH IY Z	R	*read*	R IY D
D	*dee*	D IY	S	*sea*	S IY
DH	*thee*	DH IY	SH	*she*	SH IY
DX	*body*	B AO DX IY	T	*tea*	T IY
EH	*ed*	EH D	TH	*theta*	TH EY T AH
ER	*hurt*	HH ER T	UH	*hood*	HH UH D
EY	*ate*	EY T	UW	*two*	T UW
F	*fee*	F IY	V	*vee*	V IY
G	*green*	G R IY N	W	*we*	W IY
HH	*he*	HH IY	Y	*yield*	Y IY L D
IH	*it*	IH T	Z	*zee*	Z IY
IX	*acid*	AE S IX D	ZH	*seizure*	S IY ZH ER
IY	*eat*	IY T	SIL	-	-

- Acoustic Model Topology Training

  Three states were used as the initial HMM for each phoneme. Then, a shared-state HMnet topology was obtained using an SSS training algorithm. Since the SSS training algorithm used here is based on the MDL optimization criterion, the number of shared HMM states is determined automatically by the algorithm. We call this the "MDL-SSS training algorithm." Details on MDL-SSS can be found in (Jitsuhiro et al., 2004).

- Pronunciation Dictionary

  Our pronunciation dictionary is the one used in the official "Hub and Spoke WSJ CSR Evaluation" (Kubala et al., 1994). It has been constructed by including all words from the test texts and then adding words from the WSJ0 word frequency list until 5,000 words are accumulated. Due to subject variability in reading the prompting texts, a few words have been generated that are outside the specified vocabulary.

- Language Model

  The common 5k-word bigram and trigram LM for the WSJ CSR evaluation have been generated at the MIT Lincoln Laboratory. Both are nominally closed-vocabulary grammars.

e. **Accuracy of Recognition**

The official "Hub and Spoke Paradigm for Continuous Speech Recognition Evaluation" used in 1993 for Hub 2 test data had been prepared with the participation of several institutes:

- Boston University (BU) in Boston, USA (Ostendorf and Digalakis, 1991),
- Cambridge University (CU) in Cambridge, UK (Woodland et al., 1994; Robinson et al., 1994),
- The International Computer Science Institute (ICSI) in Berkeley, USA (Morgan et al., 1994),
- France's National Center for Scientific Research (CNRS-LIMSI) in Paris, France (Gauvain et al., 1994),
- The Philips GmbH Research Laboratories in Aachen, Germany (Aubert et al., 1994).

The evaluation has been conducted under two different conditions:

- Primary condition:
  Any grammar language model or acoustic training data are allowed to be used.
- Contrast condition:
  Only the standard 5k bigram closed-vocabulary grammar and WSJ0 (7.2k utterances) are used.

Table 4.2 compares the performance of systems evaluated under these two conditions. The word error rate (WER) for contrast condition ranged from 17.7% to 8.7%, and the primary condition ranged from 9.2% to 4.9%. The lowest rate for either condition was reported by Cambridge University's HTK research group (Woodland et al., 1994). For the primary condition, evaluation has been conducted using WSJ0 and WSJ1 (SI-284) with more than 60 hours of speech. The model was a cross-word triphone, which consisted of 7,558 tied states with 10 mixture components per state. Evaluation has been done using a 5k trigram grammar.

Table 4.2. *1993 Hub and Spoke CSR evaluation on Hub 2: 5k read WSJ task (Kubala et al., 1994; Pallett et al., 1994).*

Systems	Primary WER (%)	Contrast WER (%)
BU(1)	6.7	11.6
BU(2)	5.4	10.3
BU(3)	5.8	10.8
CU(CON1)		13.5
CU(HTK2)	4.9	8.7
CU(HTK3)		12.5
ICSI		17.7
LIMSI	5.2	9.3
Phillips(1)	9.2	12.3
Phillips(2)	6.4	

For our system, we first trained the baseline triphone HMM acoustic model using the same WSJ0 and WSJ1 (SI-284) training speech data (Paul and Baker, 1992) without any knowledge of speaker's gender. We call this model "gender-independent HMM" or "GIHMM." The acoustic model is trained using the "MDL-SSS training algorithm" previously described. The total number of states is 7097, with four different numbers of Gaussian mixture components per state: 5, 10, 15, and 20. We have also incorporated additional knowledge of gender information in the baseline HMM acoustic model by training gender-dependent AMs. We call this model "gender-dependent HMM" or "GDHMM." Only embedded training has been conducted using gender-specific training data to ensure the same topology structure for all models.

Next, we obtained time-aligned state segmentation from the same WSJ training corpus and trained the HMM/BN acoustic model. The model topology, the total number of states, and the transition probabilities are all the same as those of the baseline with 7097 states. However, the HMM/BN state conditional distribution has been trained with dependence on gender information (male or female). In order to obtain the same total number of Gaussians, each gender condition (male or female) was trained corresponding to a 5-, 10-, 15-, and 20-mixture component baseline. The main functional difference between these two systems is that the HMM/BN system explores the hidden dependencies of the gender condition.

The results from the 5k Hub WSJ task are shown in Table 4.3. The best GIHMM baseline performance was obtained by the model with 7097 states with 15 and 20 mixture components per state obtaining 6.2% WER. The gender-dependent model (GDHMM) could slightly improve this performance. However, by changing the probability distribution of states to incorporate the gender information through BN (and keeping the other parameters the same), we improved recognition performance from the baseline by up to 12.9% in terms of relative WER reduction. This shows that by explicitly conditioning each Gaussian on each gender condition dependency, instead of just implicitly learning it by the EM algorithm, we can better model the overall PDF, thus achieving higher performance.

Table 4.3. *HMM/BN system performance on Hub 2: 5k read WSJ task.*

#States	#Mixtures	GIHMM baseline	GDHMM baseline	Proposed HMM/BN
7097	5	6.5	6.2	6.2
	10	6.5	5.9	5.4
	15	6.2	6.3	5.9
	20	6.2	6.2	5.9

The best results of our HMM baseline systems and the proposed HMM/BN model are illustrated in Figure 4.8, together with all systems from the "Hub and Spoke Paradigm for Continuous Speech Recognition Evaluation" for the primary condition of the WSJ Hub2-5k task. This figure shows that our systems performed competitively, and with HMM/BN we have been able to improve our position to roughly the third ranking one.

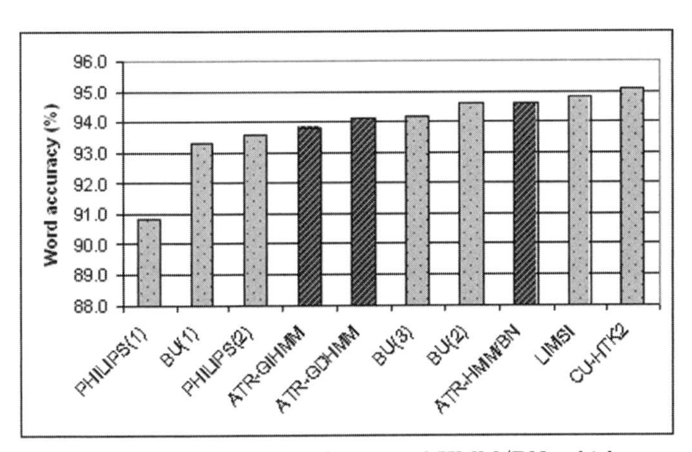

Fig. 4.8. Recognition accuracy rates of proposed HMM/BN, which are comparable with those of other systems from the "Hub and Spoke Paradigm for Continuous Speech Recognition Evaluation" for primary condition of WSJ Hub2-5k task.

Incorporating Background Noise Information

When speech is contaminated by background noise, feature vectors change their distributions, and this change depends on the noise type as well as on the signal-to-noise ratio (SNR) (Markov and Nakamura, 2003). By applying GFIKS, we can incorporate both noise type and SNR information at each HMM state and then express their dependencies as defined below.

a. Causal Relationship and Inference

The causal relationship between HMM state Q, observation vector X, and additional knowledge sources of noise type N and SNR value S may be described with the BN topology shown in Figure 4.9.

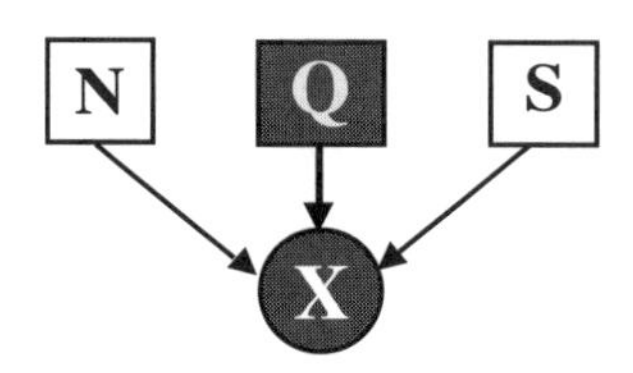

Fig. 4.9. BN topology structure describing the conditional relationship between HMM state Q, observation vector X, and additional knowledge sources of noise type N and SNR value S.

Following Eq. (4.2), the HMM state PDF is the BN joint probability model, which is expressed as

$$P(X, N, S, Q) = P(X|N, S, Q)P(N)P(S)P(Q), \tag{4.13}$$

where the model depends on noisy type N and SNR value S. When N and S are observable during inference, the HMM state output probability is simply

$$p(x_t|n_m, s_n, q_j) = P(X = x_t|N = n_m, S = s_n, Q = q_j). \tag{4.14}$$

However, assuming N and S to be hidden (unknown) during recognition, the state output probability is then calculated by marginalization over N and S:

$$p(x_t|q_j) = \sum_{m=1}^{M_N} \sum_{n=1}^{M_S} p(n_m)p(s_n)p(x_t|n_m, s_n, q_j), \tag{4.15}$$

where, for simplicity, we use the x_t, q_j, n_m and s_n notations instead of $\langle X = x_t \rangle$, $\langle Q = q_j \rangle$, $\langle N = n_m \rangle$ and $\langle S = s_n \rangle$, respectively. $p(n_m)$ is the probability that state q_j has noisy type n_m, $p(s_n)$ is the probability that state q_j has SNR value s_n, and $p(x_t|n_m, s_n, q_j)$ is the probability of observation x_t given that we are in state q_j having noise type n_m and SNR value s_n. Here, we can see that Eq. (4.15) is equivalent to the state output probability of the conventional HMM of Eq. (2.43) if we treat the term $p(n_m|q_j)p(s_n|q_j)$ as a mixture weight coefficient for the Gaussian component $P(X|n_m, s_n, q_j)$.

b. Enhancing Model Reliability

Different real-world noises N are selected over a range of SNR S values. Since the total number of corresponding Gaussian mixture components will also be limited, no clustering technique needs to be performed here.

c. Training and Recognition Issues

The training procedure is based on the algorithm described in Section 4.1.4. Since all variables, including triphone state Q, noise type N, SNR value S, and feature variable X, are observable during training, only simple ML parameter estimation is applied to the training of the state BN in step 3 of the algorithm.
Recognition is performed using the existing HMM-based decoders without any modification. The HMM state output probability is calculated using Eq. (4.15).

d. Experimental Set-Up

The experimental set-up followed closely the evaluation scenario used for the well-known AURORA2 evaluation task (Hirsch and Pearce, 2000). This is a noisy-speech speaker-independent digits task, generated by Ericsson Eurolab as a contribution to the ETSI STQ-AURORA DSR Working Group. The source speech of this database is TIDigits, consisting of a connected-digits task using the following eleven words: "zero," "oh," "one," "two," "three," "four," "five," "six," "seven," "eight," and "nine," spoken by American English talkers (a brief explanation can also be found in Appendix A.1).

The detailed parameter set-up is defined as follows:

- Front-End Parameters

  For feature extraction, we used a sampling frequency of 16 kHz, a frame length of a 20-ms Hamming window, a frame shift of 10 ms, and standard 39-dimensional feature parameters consisting of 12 MFCC coefficients plus power, as well as Δ and $\Delta\Delta$ MFCC.

- Unit Set

  No phoneme set has been used here, since only eleven words can be recognized. Each word is modeled by each HMM.

- Acoustic Model Topology Training

  Since this is based on a word model, 16-state HMMs have been used for each word. No special state-tying training algorithm has been performed here.

- Pronunciation Dictionary

  No pronunciation dictionary has been used here, since it is only a word-based HMM of digits.

- Language Model

  Furthermore, no LM has been used here, since there is no syntactical or semantical restriction on the sequence of digits.

e. Accuracy of Recognition

Prior work by Markov and Nakamura (2003) describes some experimental results on incorporating noise information in the HMM state. A summary of these experiments is presented here.

The primary interest is to compare the HMM/BN system with a multi-condition trained HMM system. The training set consists of 8440 utterances (about 4.167 hours of speech time), which are spoken by 55 female and 55 male speakers. A selection of 8 different real-world noises is added over a range of SNR ratios, which consists of four different noises (subway, babble, car, exhibition hall), and four SNR values (20 dB, 15 dB, 10 dB, 5 dB).

Each word in the multi-conditional HMM baseline system is modeled by a 16-state HMM with 3 mixtures per state. Only the silence model uses 3 states with 6 mixtures per state. Using the same training set, we have obtained time-aligned state segmentation. The model topology, the total number of states, and the transition probabilities are all the same as those of the HMM baseline. The training set has been divided by noise type and SNR value, and the HMM/BN state conditional distribution has been trained for each condition separately.

Test sets A and B were used here, where test set A contains the same noise types used in multi-condition training, while test set B contains four different noises: restaurant, street, airport and train station. In both test sets, two SNR values, 0 dB and -5 dB, are added.

Table 4.4 summarizes the recognition results for both test sets A and B. As discussed in (Markov and Nakamura, 2003), the HMM/BN model performance was much higher than the baseline HMM system for the closed noise condition test (test set A), especially for the low SNR conditions. This shows that by incorporating the noise type and SNR value information, the state Gaussian mixtures of the HMM/BN model could model the complex distribution of multiple-noise and SNR conditions in a better way. On average, this yielded 36.4% relative improvement. However, in the mismatch noise condition (test set B), performance degradation occurred. One reason for this was the fact that the state Gaussian mixtures of the HMM/BN model have no knowledge available on the new noise types. On the other hand, the state Gaussian mixtures of the HMM baseline clearly do not model very well the complex distribution from multiple-noise and SNR conditions, which, however, makes it easier to generalize over unseen data.

Table 4.4. *Recognition accuracy rates (%) for proposed HMM/BN on AURORA2 task.*

SNR	Test set A		Test set B	
	HMM	HMM/BN	HMM	HMM/BN
Clean	98.54	98.83	98.54	98.83
20 dB	97.52	98.12	96.96	97.26
15 dB	96.94	97.65	95.38	95.05
10 dB	94.59	96.04	92.58	90.27
5 dB	87.51	91.70	83.50	78.00
0 dB	59.84	76.11	58.91	48.70
-5 dB	23.46	35.79	23.86	3.18

In addition, we compared the performances of three different systems: HMM, DBN (Bilmes et al., 2001), and the proposed HMM/BN (Figure 4.10). As can be seen by the results, the proposed HMM/BN performs even better than DBN, especially at low SNRs.More detailed experiments can be found in (Markov and Nakamura, 2003).

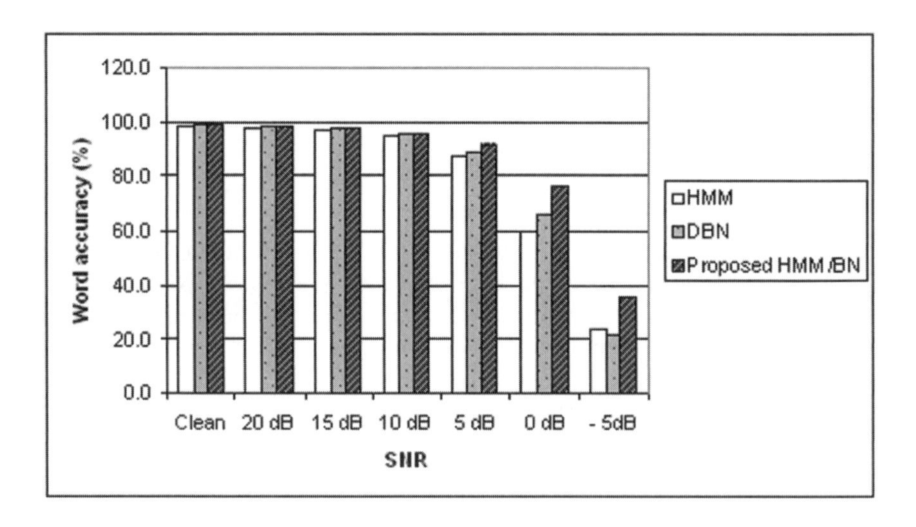

Fig. 4.10. Comparison of different systems: HMM, DBN (Bilmes et al., 2001), and proposed HMM/BN

Incorporating Wide-phonetic Context Information

As discussed in Section 3.3.1, by incorporating a wider context than the triphone, more than just one preceding and one following phonetic context may be taken into account. It may be possible to improve the acoustic capability to handle the coarticulation effects that exist in everyday conversational speech. In this work, we attempted to incorporate a pentaphone context having the form $/a^{--}, a^{-}, a, a^{+}, a^{++}/$.

a. Causal Relationship and Inference

Based on our GFIKS approach, we have extended our conventional HMM with triphone $/a^{-}, a, a^{+}/$ to the pentaphone $/a^{--}, a^{-}, a, a^{+}, a^{++}/$. This has been achieved by adding second preceding and succeeding contexts, C_L ($/a^{--}/$) and C_R ($/a^{++}/$), to the triphone state PDF using the BN.

One possible implementation of this approach is the assumption that the added preceding and following contexts mainly affect the outer states of the triphone HMM model, so that only the left and right states have additional knowledge. The left, center, and right state output probability distributions can be represented by three different BN topologies as shown in Figure 4.11(a), (b) and (c), respectively.

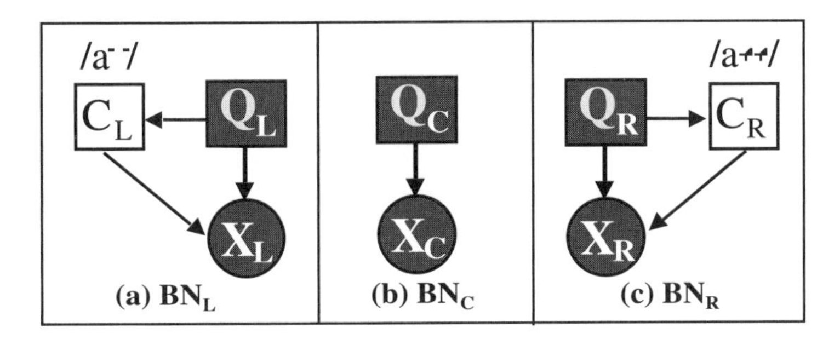

Fig. 4.11. BN topologies of the left state (a), center state (b), and right state (c) of LR-HMM/BN for modeling a pentaphone context $/a^{--}, a^{-}, a, a^{+}, a^{++}/$.

As can be seen, only BN_L and BN_R have an additional discrete variable C_L and C_R associated with the second preceding and following contexts, respectively. BN_C does not have any additional context variable. We call this model LR-HMM/BN.

The state PDF of the pentaphone HMM/BN model is the BN joint probability model, which is expressed as:

$$P(X, C, Q) = P(X|C, Q)P(C|Q)P(Q), \tag{4.16}$$

where the model depends on the second preceding or succeeding context C. When C is observable, the left/right state output probability is simply

$$p(x_t|c_m, q_j) = P(X = x_t|C = c_m, Q = q_j). \tag{4.17}$$

However, since the second preceding/following context C (C_L or C_R) is assumed hidden during recognition, the left/right state output probability is then calculated by marginalization over C:

$$p(x_t|q_j) = \sum_{m=1}^{M} p(c_m|q_j)p(x_t|c_m, q_j), \tag{4.18}$$

where, for simplicity, we use the x_t, q_j, and c_m notations instead of $\langle X = x_t \rangle$, $\langle Q = q_j \rangle$, and $\langle C = c_m \rangle$, respectively. $p(c_m|q_j)$ is the probability that state q_j has the second preceding/following contexts c_m, and $p(x_t|c_m, q_j)$ is the probability of observation x_t given that we are in state q_j having the second preceding/following contexts c_m. Here, we can see that Eq. (4.18) is equivalent to the state output probability of the conventional HMM of Eq. (2.43) if we treat the term $p(c_m|q_j)$ as a mixture weight coefficient for the Gaussian component $P(X|c_m, q_j)$.

We then incorporated the wide-context dependencies into the center state of the triphone HMM model. The state BN topologies for this case are shown in Figure 4.12. BN_L and BN_R are the same as before, while BN_C has two additional context variables: the second preceding (C_L) and the second following (C_R) contexts. Since all states have wide-context variables, we call this model LRC-HMM/BN.

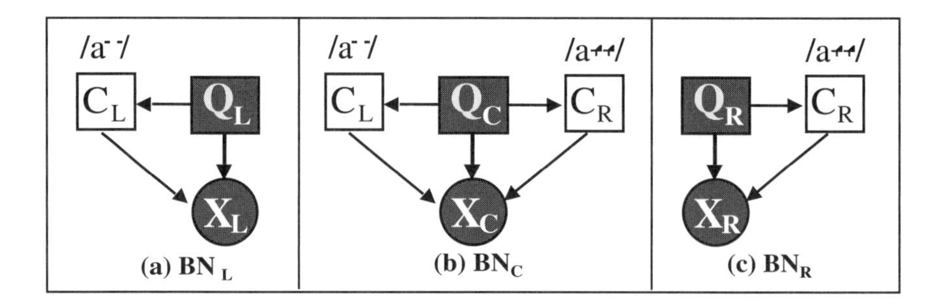

Fig. 4.12. BN topologies of the left state (a), center state (b), and right state (c) of LRC-HMM/BN, for modeling a pentaphone context $/a^{--}, a^{-}, a, a^{+}, a^{++}/$.

The output probability for the left/right state is obtained as in LR-HMM/BN. Here, the center state output probability is obtained from BN_C, assuming also that both additional knowledge C_L and C_R are hidden during recognition and take N_L and N_R values:

$$p(x_t|q_j) = \sum_{l=1}^{M_L} \sum_{r=1}^{M_R} p(c_l|q_j)P(c_r|q_j)p(x_t|c_l, c_r, q_j), \qquad (4.19)$$

where, we use x_t, q_j, c_l, and c_r notations instead of $\langle X = x_t \rangle$, $\langle Q = q_j \rangle$, $\langle C_L = c_l \rangle$, and $\langle C_R = c_r \rangle$, respectively. $p(c_l|Q)p(c_r|q_j)$ are the probabilities that the center state q_j has the second preceding and following contexts (c_l and c_r), and $p(x_t|c_l, c_r, q_j)$ is the probability of observation x_t given that we are in the center state q_j having the second preceding and following contexts, c_l and c_r, respectively. Here, we can see that Eq. (4.19) is also equivalent to the state output probability of the conventional HMM of Eq. (2.43) if we treat the term $p(c_l|q_j)P(c_r|q_j)$ as a mixture weight coefficient for the Gaussian component $P(X|c_l, c_r, q_j)$.

b. Enhancing Model Reliability

The observation space modeling by BN, where a different value of the second following context C_R corresponds to a different Gaussian is shown in Figure 4.13. Here, for this additional context, we used both knowledge-based phoneme classes and data-driven clustering techniques.

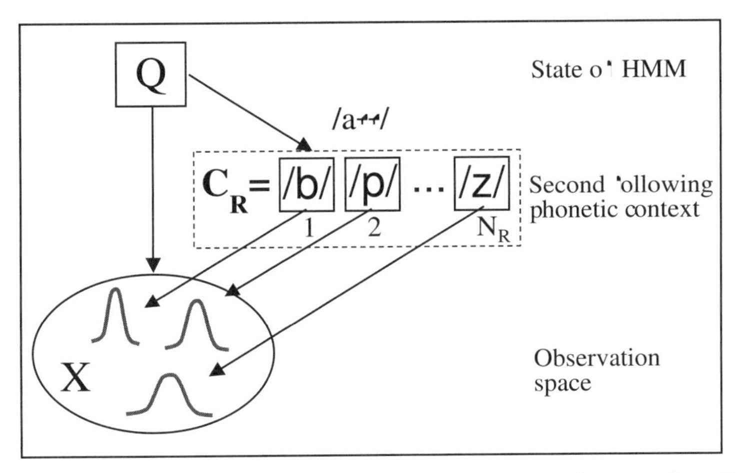

Fig. 4.13. Observation space modeling by BN, where a different value of second following context C_R corresponds to a different Gaussian.

- Knowledge-based Approach

  In this method, the specific knowledge of the contextual unit is explicitly used to guide the classification procedure (Huang et al., 2001). For example, if our additional knowledge K_i represents phoneme contexts, then we can use major distinctions in the manner of articulation. Many phonemes having the same location of articulation tend to have similar effects on the neighboring phonemes. For example, /b/ and /p/ have similar effects on the following vowel, as do /n/ and /m/. Here, in order to reduce the parameter size, we group the phoneme contexts based on major distinctions in the manner of articulation. Table 4.5 shows an example of knowledge-based phoneme classes adapted from the classification in (Odell, 1995).

Table 4.5. *Knowledge-based phoneme classes based on manner of articulation.*

Classes	Phonemes
Plosives	b, d, g, k, p, t
Nasal	m, n, ng
Fricatives	ch, dh, f, jh, s, sh, th, v, z, zh
Liquid	hh, l, r, w, y
Vowels	ih, ix, iy, eh, ey, aa, ae, aw, axr, ay, er, ao, ow, oy, uh, ah, ax, uw

By also considering the amount of training data, each of these terminal nodes is divided into more detailed nodes, such as plosive bilabials, plosive velars, and fricative glottals. Based on this tree, we can cluster N (N_L or N_R) second preceding/following contexts into L classes where $L < N$ as shown in Figure 4.14.

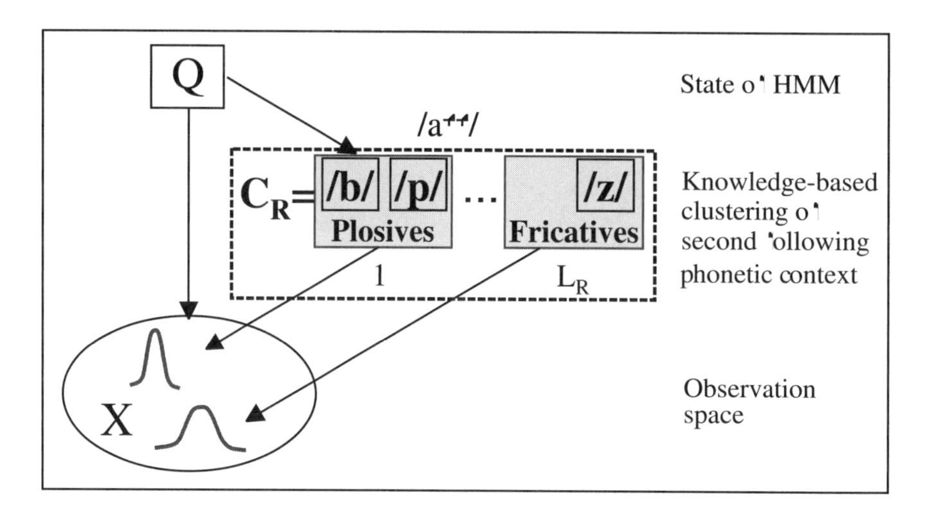

Fig. 4.14. Knowledge-based phoneme classes of the observation space.

- Data-driven Approach

  Data-driven clustering is also a common approach to parameter tying. Instead of clustering the data based on specific knowledge, they are clustered based on some similarity measure, regardless of the phonetic context they represent (Friedman and Goldszmidt, 1998). Initially, each Gaussian is placed in a separate cluster, and then the clusters that would form the smallest resulting cluster when combined are merged. The distance metric is determined by the Euclidean distance between the Gaussian means as shown in Figure 4.15.

  This process is repeated until the total number of clusters falls below a certain threshold. With this clustering technique, we can set-up any total number of Gaussian components so that it corresponds to the averaged fixed number of mixture components per state, i.e., L mixture components, where $L < N$ as shown in Figure 4.16.

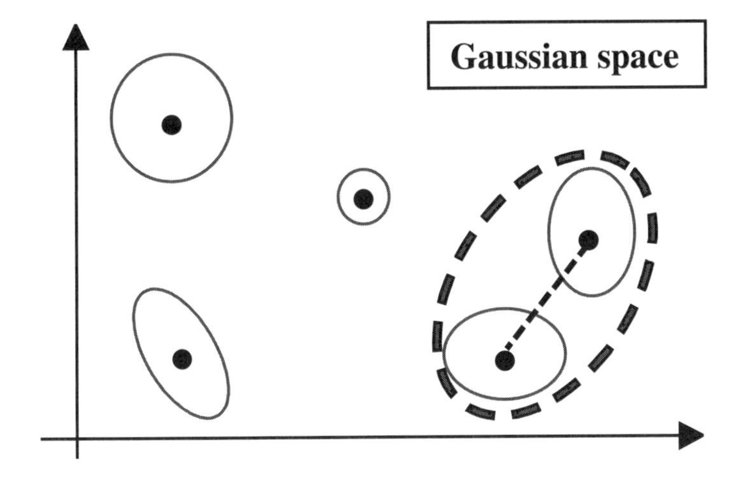

Fig. 4.15. Determining distance metric by Euclidean distance.

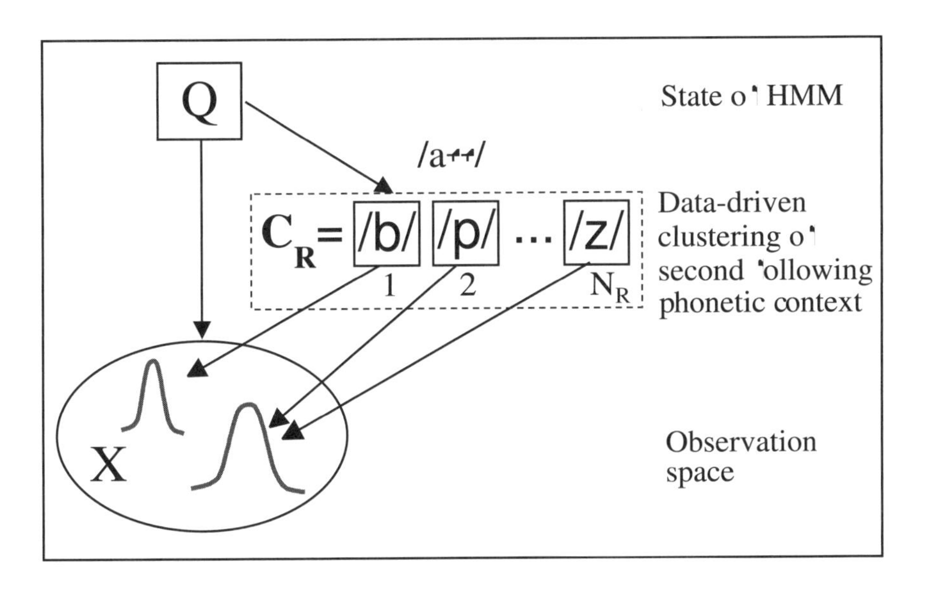

Fig. 4.16. Data-driven phoneme classes of observation space.

c. Training and Recognition Issues

The training procedure is based on the algorithm described in Section 4.1.4. Since all variables, including triphone state Q, second preceding (C_L) context, second following (C_R) context, and feature variable X, are

observable during training, only simple ML parameter estimation is applied to the training of the state BN in step 3 of the algorithm.
We perform recognition using the existing triphone HMM-based decoders without any modification. The state output probability is calculated using Eqs. (4.18) and (4.19).

d. Experimental Set-Up

These experiments were conducted on the ATR basic travel expression corpus (BTEC) task. The ATR basic parameter set-up can be described as follows:

- Front-End Parameters

  Speech data are processed with 16 kHz sampling frequency, a frame length of a 20-ms Hamming window, a frame shift of 10 ms, and 25-dimensional feature parameters consisting of 12-order MFCCs, Δ MFCCs and Δ log power.

- Phoneme Set

  The phoneme set used here consists of 43 English phonemes plus one silence (SIL) as listed in Table 4.1.

- Acoustic Model Topology Training

  Acoustic models have been trained using the "MDL-SSS training algorithm."

- Pronunciation Dictionary

  The pronunciation dictionary used here consists of about 37 k words and is based on US-accented pronunciations.

- Language Model

  We have used both bigram and trigram language models that have been trained on about 150,000 travel-related sentences.

e. Accuracy of Recognition

We have used SI-284 WSJ training speech data to train our baseline triphone HMM acoustic model (see Section A.3). As our standard mechanism, three states have been used in the initial HMM for each phoneme. Then, a shared-state HMnet topology was obtained using an SSS training algorithm. Since incorporating wide-phonetic context information increases the number of parameters, we set up the MDL parameter to reduce the number of states, i.e., 1,144. Four different numbers of Gaussian mixture components per state have been generated: 5, 10, 15, and 20. Each Gaussian distribution has a diagonal-covariance matrix.

The performances of the models have been tested on the ATR BTEC (Takezawa et al., 2002, a brief explanation can also be found in Appendix A.4), which is quite different from the training corpus. In this study, we randomly selected 200 utterances from 4,080 utterances spoken by 40 different speakers (20 Males, 20 Females). The best baseline HMM performance was 87.98% word accuracy, obtained by a triphone HMM with 15 Gaussians per state.

Using the same database corpus, we obtained time-aligned state segmentation. First, we evaluated the hybrid pentaphone LR-HMM/BN and trained BN_L/BN_R with second preceding/following contexts as additional discrete variables. The center state BN_C is equivalent to the standard HMM state PDF modeled as a mixture of Gaussians. Thus, as a center state of the HMM/BN model, we have used the five corresponding component mixture states from the baseline acoustic model. The HMM/BN state topology, the total number of states, and the transition probabilities are all the same as those of the baseline.

The initial HMM/BN model used a 44-phoneme context set for C ($C = c_1, c_2, ..., c_{44}$). During training, some phoneme contexts c_n did not exist due to grammatical rules or were unseen in the training data, which after training resulted in about 30 Gaussians on average per left/right state. Since the center-state parameters remain the same as those of the baseline triphone 5-mixture-component HMM, the final hybrid LR-HMM/BN model has about 24 mixtures per state (on average).

Then, as described in part (b.)"Enhancing Model Reliability" above, we have reduced the 44-phoneme set to 30, 20, 10, and 6 classes by using knowledge-based phoneme clustering. Keeping the center state at five Gaussians per state resulted in hybrid LR-HMM/BN models with 18, 13, 8, and 5 component mixtures on average, respectively. The results of the pentaphone LR-HMM/BN with different kinds of phoneme class sets are shown in Figure 4.17. For comparison, we have included the HMM triphone baseline with the 15 component mixtures, which showed optimum performance.

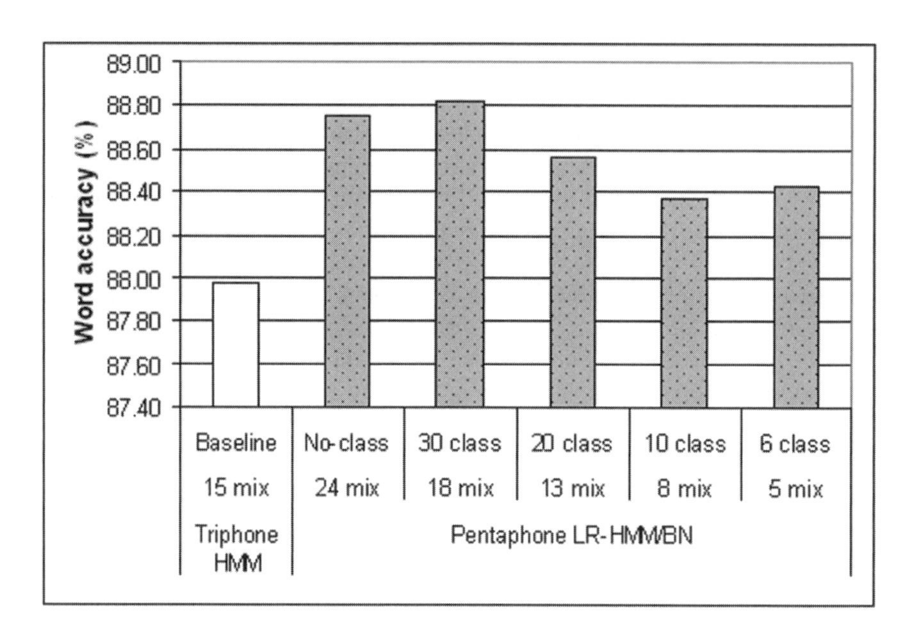

Fig. 4.17. Recognition accuracy rates of pentaphone LR-HMM/BN using knowledge-based second preceding and following context clustering.

We have also evaluated the hybrid pentaphone LRC-HMM/BN model and trained the BN_C with both second preceding and following contexts as additional discrete variables. The left and right states (BN_L and BN_R) are the same as in the hybrid pentaphone LR-HMM/BN. The HMM/BN state topology, the total number of states, and the transition probability are also those of the baseline. The initial HMM/BN model used a 44-phoneme context set for C ($C = c_1, c_2, ..., c_{44}$). During training, some phoneme contexts c_n did not exist due to grammatical rules or were unseen in the training data, which after training resulted in about 412 Gaussians on average per center state and 30 Gaussians on average per left/right state. The average for the final hybrid pentaphone LRC-HMM/BN model was about 142 mixtures per state. To reduce the number of Gaussians, we clustered the 44-phoneme-context set into 30, 20, 10, and 6 classes using knowledge-based phoneme clustering. As a result, the hybrid pentaphone LRC-HMM/BN models had 108, 70, 29, and 13 component mixtures, respectively.

The results of the pentaphone LRC-HMM/BN with different kinds of phoneme class sets are shown in Figure 4.18. By changing only the probability distribution of states to incorporate a wider phonetic context through BN (and keeping the other parameters the same), the recognition performance could be improved. The pentaphone LR-HMM/BN with 30 classes is the best, and further reducing the number of parameters degrades performance. Nevertheless, even the worst performance is still better than the baseline. The pentaphone LRC-HMM/BN with a 44-phoneme set (142 mixtures per state) performed only slightly better than the HMM baseline due to the huge number of parameters. By reducing the number of Gaussians, the resulting performance can be improved from 88.05% to 88.82%. This best performance of the pentaphone LRC-HMM/BN is obtained with 10 classes (29 Gaussians per state). For the optimal size of C_L and C_R using the knowledge-based phoneme clustering, both LRC-HMM/BN and LR-HMM/BN models achieved similar performance.

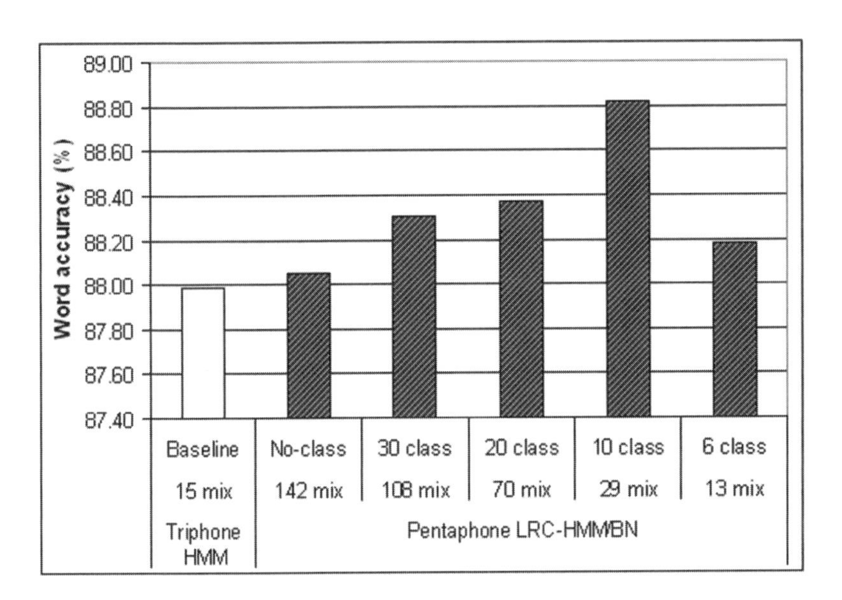

Fig. 4.18. Recognition accuracy rates of pentaphone LRC-HMM/BN using knowledge-based second preceding and following context clustering.

In order to compare the pentaphone HMM/BN model and the baseline having the same total number of Gaussians, we used data-driven clustering to reduce the size of the initial HMM/BN model to correspond to a 5-, 10-, 15-, and 20-mixture component baseline. The center state of the pentaphone LR-HMM/BN also had the corresponding mixture component size.

The results of the triphone HMM baseline, the pentaphone LR-HMM/BN, and the pentaphone LRC-HMM/BN are shown in Figure 4.19. It can be seen that within the same number of parameters, both types of pentaphone HMM/BN outperform the baseline. The best performance of the pentaphone LR-HMM/BN is obtained with 15 Gaussian mixtures, which provides about a 9% reduction in relative WER, while the best performance of the pentaphone LRC-HMM/BN is obtained with 20 Gaussian mixtures, which gives about a 10% reduction in relative WER. Both differences are significant at the 5% level calculated using the Sign test (Hays, 1988). A brief explanation of significant hypothesis can also be found in Appendix D. On average, both the LRC-HMM/BN and LR-HMM/BN models also achieved similar performances as before, indicating that knowledge-based and data-driven clustering techniques are equally efficient in reducing the number of Gaussian components.

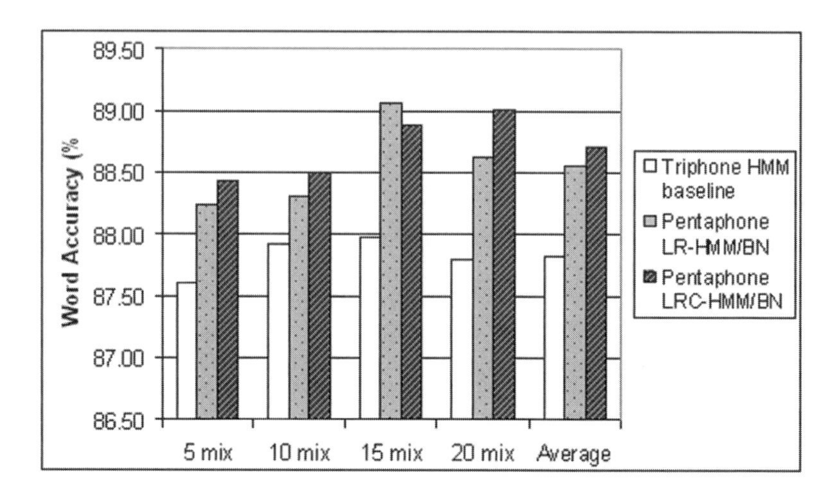

Fig. 4.19. Recognition accuracy rates of pentaphone LR-HMM/BN and LRC-HMM/BN using data-driven Gaussian clustering.

Work by other researchers has indicated that a model with a varied number of mixture components often outperforms a model with a fixed number of mixture components, when both models have almost the same total number of Gaussians (Valtchev et al., 1997). To confirm that the superior performance of our proposed models is not caused by this effect, we conducted additional experiments using a triphone HMM model with a varied number of mixture components per state. This model was trained by simply assigning the number of mixture components per state depending on the amount of training data for that state. The LR-HMM/BN had a fixed number of mixture components per state trained by applying data-driven clustering for each state.

With both models having approximately 15 mixture components per state, their performances were compared with the baseline and the previous pentaphone HMM/BN models, and the results are shown in Figure 4.20. The performance of the LR-HMM/BN with a fixed number is still better than that achieved by the triphone models with a varied number of mixture components. This indicates that the coarticulation variability is higher than the variability of most other factors. Thus, by explicitly conditioning each Gaussian on such pentaphone-context dependency, instead of just implicitly learning it using the EM algorithm, we can better model the overall PDF, thus achieving an improvement in performance.

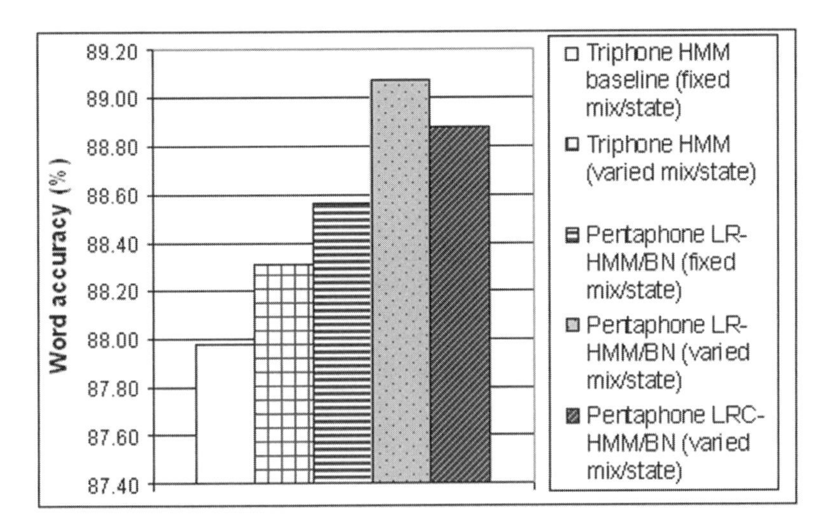

Fig. 4.20. Comparing recognition accuracy rates of triphone HMM and pentaphone HMM/BN models with a fixed and a varied number of mixture components per state, but having the same 15 mixture components per state on average.

Incorporating Multiple Knowledge: Accent, Gender and Wide-phonetic Context Information

a. Causal Relationship and Inference

Using the GFIKS framework, we may also further extend the pentaphone BN with other additional knowledge variables, such as gender or accent information. To simplify the topology, we first set-up the incorporation of two additional knowledge, C_L and C_R, in the same way for each triphone HMM state Q (left, center and right) as shown in Figure 4.21. We call this the full LRC-HMM/BN, or fLRC-HMM/BN, topology.

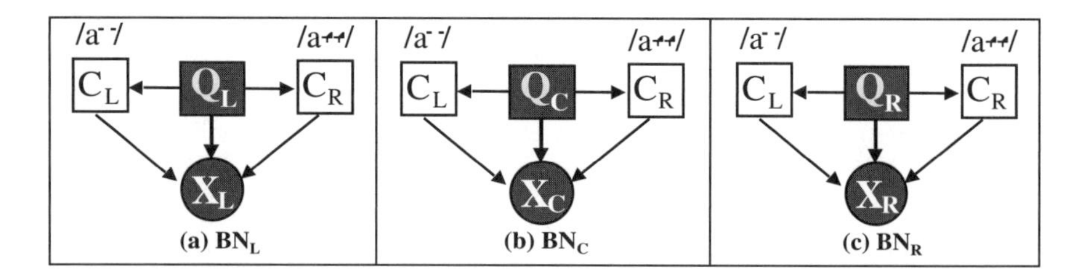

Fig. 4.21. Topology of fLRC-HMM/BN for modeling a pentaphone context $/a^{--}, a^{-}, a, a^{+}, a^{++}/$, where state PDF has additional variables C_L and C_R representing the second preceding and following contexts, respectively.

Each HMM state PDF is currently represented by the BN joint probability, which according to Eq. (4.2) can be decomposed as

$$\begin{aligned} &P(X, C_L, C_R, Q) \\ &= P(X|C_L, C_R, Q)P(C_L|Q)P(C_R|Q)P(Q), \end{aligned} \tag{4.20}$$

where X depends on both second preceding context C_L and second following context C_R. Since X is continuous and C_L, C_R, and Q are discrete variables, $P(X|C_L, C_R, Q)$ is modeled with a Gaussian function, and both $P(C_L|Q)$ and $P(C_R|Q)$ are represented by a CPT.

The state output probability can be obtained from $P(X|C_L, C_R, Q)$, and assuming that the additional context variables, C_L and C_R, cannot be observed (are hidden) during recognition, as in Eq. (4.3),

$$p(x_t|q_j) = \sum_{l=1}^{M_L} \sum_{r=1}^{M_R} p(c_l|q_j)p(c_r|q_j)p(x_t|c_l, c_r, q_j), \tag{4.21}$$

which is equivalent to the state output probability of the conventional HMM in Eq. (2.43), if we treat term $p(c_l|q_j)p(c_r|q_j)$ as a mixture weight coefficient for the Gaussian component, $P(X|c_l, c_r, q_j)$. Consequently, here a Gaussian PDF is trained for all combinations of c_l, c_r, q_j.

We can now further extend the pentaphone BN with gender or accent information using this framework. Figure 4.22 describes several examples of conditional relationship structures among triphone HMM state Q, observation data X, the two additional variables, C_L and C_R, and the gender, G, or accent, A, variables. The BN topology becomes that illustrated in Figure 4.22(a) by extending fLRC-HMM/BN with an additional variable of gender G, and this is denoted as fLRCG-HMM/BN. The BN topology becomes the one in Figure 4.22(b) by extending fLRC-HMM/BN with the additional accent variable A, and this is called fLRCA-HMM/BN. The BN topology in Figure 4.22(c) is extended with both accent and gender variables, and this is called fLRCAG-HMM/BN.

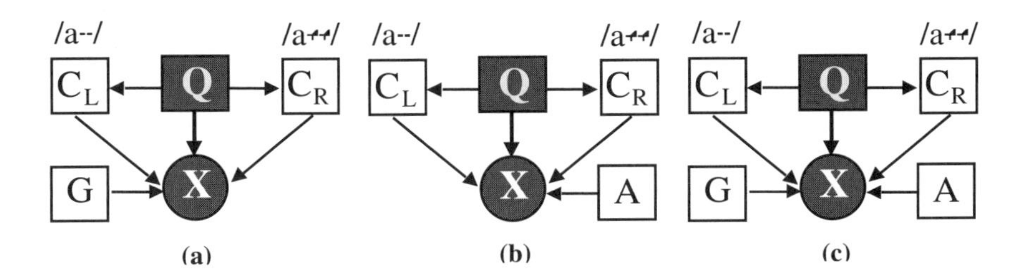

Fig. 4.22. (a) fLRCG-HMM/BN topology with additional knowledge G, C_L and C_R, (b) fLRCA-HMM/BN topology with additional variables A, C_L, and C_R, and (c) fLRCAG-HMM/BN topology with additional knowledge A, G, C_L, and C_R.

Each HMM state PDF for the fLRCAG-HMM/BN example (see Figure 4.22(c)) is expressed as

$$\begin{aligned} & P(X, C_L, C_R, Q, A, G) \\ & = P(X|C_L, C_R, Q, A, G) P(C_L|Q) P(C_R|Q) \\ & \quad P(Q) P(A) P(G), \end{aligned} \tag{4.22}$$

where X depends on accent A, gender G, the second preceding context, C_L, and the second following context, C_R. The state output probability can also be obtained from $P(X|C_L, C_R, Q, A, G)$ in a similar way to that in Eq. (4.21):

$$\begin{aligned} p(x_t|q_j) = & \sum_{n=1}^{M_A} \sum_{m=1}^{M_G} \sum_{l=1}^{M_L} \sum_{r=1}^{M_R} p(a_n) p(g_m) p(c_l|q_j) p(c_r|q_j) \\ & p(x_t|c_l, c_r, q_j, a_n, g_m). \end{aligned} \tag{4.23}$$

Here, we also treat the term, $p(a_n)p(g_m)p(c_l|q_j)p(c_r|q_j)$, as a mixture weight coefficient for the Gaussian component, $P(X|c_l, c_r, q_j, a_n, g_m)$, so that each Gaussian PDF is trained for each combination of c_l, c_r, q_j, a_n, g_m.

b. Enhancing Model Reliability

If the amount of training data is not enough to obtain a reliable estimate of the increased model parameters, the overall performance may degrade significantly. Any type of clustering technique as described in Section 4.3.1, i.e., knowledge-based or data-driven clustering, can also be applied here.

c. Training and Recognition Issues

The training procedure is also based on the algorithm described in Section 4.1.4. Since all variables, including triphone state Q, accent A, gender G, second preceding (C_L) context, second following (C_R) context, and feature variable X are observable during training, only simple ML parameter estimation is applied to the training of the state BN in step 3 of the algorithm.

Recognition has been performed using existing triphone HMM based decoders without modification. The state output probability has been calculated using Eqs. (4.21) and (4.23).

d. Experimental Set-Up

These experiments were conducted on accented speech of the ATR BTEC task. The ATR basic parameter set-up is similar to the one described in Section 4.3.1.

e. Accuracy of Recognition

The proposed pentaphone models have been trained using the same amount of training data for all accent data labeled with phoneme-class context variables. The model state topology, the total number of states, and the transition probabilities were all identical to those of the triphone HMM baseline. Therefore, they all had similar complexity in terms of the number of parameters. The main difference is only in the probability distribution of states where each Gaussian has been explicitly conditioned on C_L or C_R. All Gaussian components in the HMM baseline, in contrast, have been learned implicitly by the EM algorithm, without any "meaningful" interpretation of the mixture index.

Some phoneme context classes of C_L or C_R did not exist due to grammatical rules or did not appear in the training data, which after training resulted in about 50 Gaussians per state on average. We used a data-driven clustering technique and reduced the size of the pentaphone models to correspond to 5, 10, 15, and 20 mixture components per state; this was done to avoid unreliably estimated parameters and to compare in a valid way the performance with that of the baseline system by having exactly the same total number of Gaussians.

The performances for the models having 5, 10, and 20 mixture components per state are shown in Figure 4.23. For the case of 5 mixture components per state, the triphone baseline without any additional knowledge achieved 83.60% word accuracy. However, for the accent-gender-dependent models, it decreased to 82.11% word accuracy. This may be due to the size of the

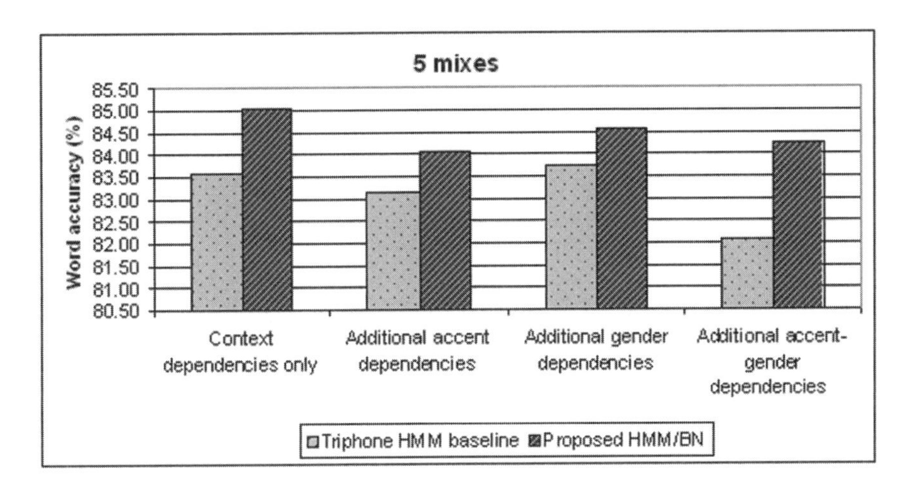

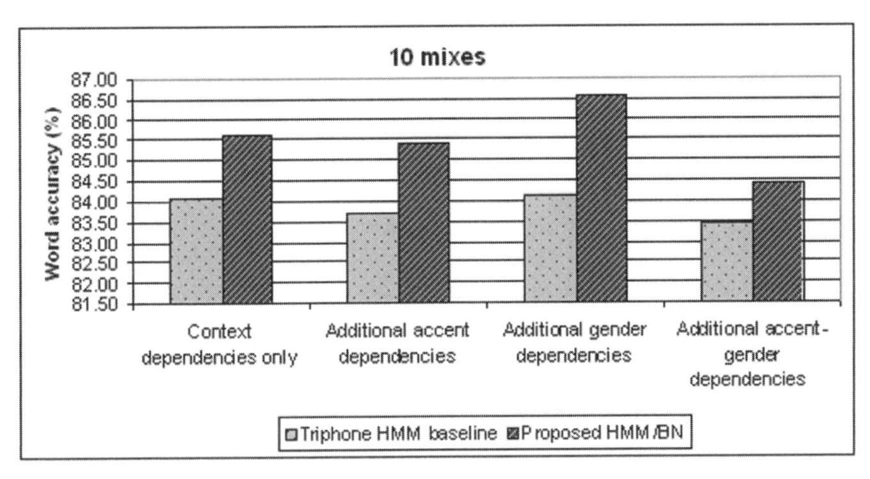

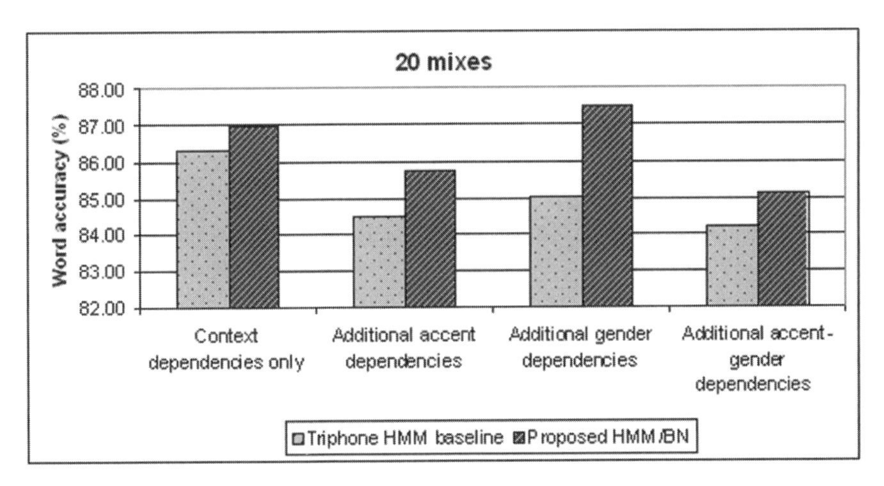

Fig. 4.23. Recognition accuracy rates of proposed HMM/BN models having identical numbers of 5, 10, and 20 mixture components per state.

training data, which is much smaller compared to the other baseline models. By changing only the probability distribution of states to incorporate various types of knowledge sources through the BN, we obtained significantly improved recognition of 85.03% word accuracy. Overall, through using different mixture components per state, the results show that, the proposed pentaphone models consistently outperform the standard HMM baseline. The differences are significant at the 5% level calculated using the Sign test (see Appendix D).

To investigate the effect of using fLRC-HMM/BN in more detail, we have evaluated it on a test set of matching accents, where the test data consisted of 200 randomly selected utterances from each accent type (US and AUS). The results obtained with models of different numbers of mixture components are summarized in Table 4.6. It can be seen that the proposed pentaphone models always performed better than the baseline with the same number of parameters. The best performance for the US pentaphone HMM/BN was obtained with 10 Gaussian mixtures, which resulted in a relative reduction in WER of about 8%, and the best performance for the AUS pentaphone was obtained with 20 Gaussian mixtures, which resulted in a relative reduction in WER of about 11%.

Table 4.6. *Recognition accuracy rates (%) for proposed pentaphone HMM/BN model using fLRC-HMM/BN (see Figure 4.22) on a test set of matching accents with different numbers of mixture components.*

	US accent		**AUS accent**	
Mixture number	**Triphn baseline**	**Proposed HMM/BN**	**Triphn baseline**	**Proposed HMM/BN**
5 mix	84.30	85.19	82.33	84.24
10 mix	84.66	85.91	82.21	84.12
15 mix	84.78	85.55	83.46	84.18
20 mix	85.25	85.67	82.63	84.60

We have also evaluated the performance of these pentaphone models on a test set of mismatched accents, e.g., the model trained on US speech has been tested on the AUS speech test data and vice versa. The results obtained using the models with 15 mixture components are summarized in Table 4.7. The results from evaluating matching accents are also included to enable easy comparison. We can see that the pentaphone model on mismatched accents still consistently outperforms the standard HMM triphone model.

Table 4.7. *Recognition accuracy rates (%) for proposed pentaphone HMM/BN model using fLRC-HMM/BN (see Figure 4.22) on a test set of mismatched accents with 15 mixture components.*

Accented test set	US accent Triphn baseline	US accent Proposed HMM/BN	AUS accent Triphn baseline	AUS accent Proposed HMM/BN
US test	84.78	85.55	75.22	76.96
AUS test	64.78	65.43	83.46	84.18

We have conducted additional experiments on a conventional pentaphone HMM model with 2,202 states, which has been trained from scratch using MDL-SSS, to investigate whether the superior performance of our proposed models is due mainly to the wide-phonetic context. In order to avoid decoding complexity, the model has been implemented by rescoring the N-best list generated from a standard and unmodified triphone ASR system, as we did in our previous studies (Sakti et al., 2006). The results for all types of models with five mixture components per state are displayed in Figure 4.24.

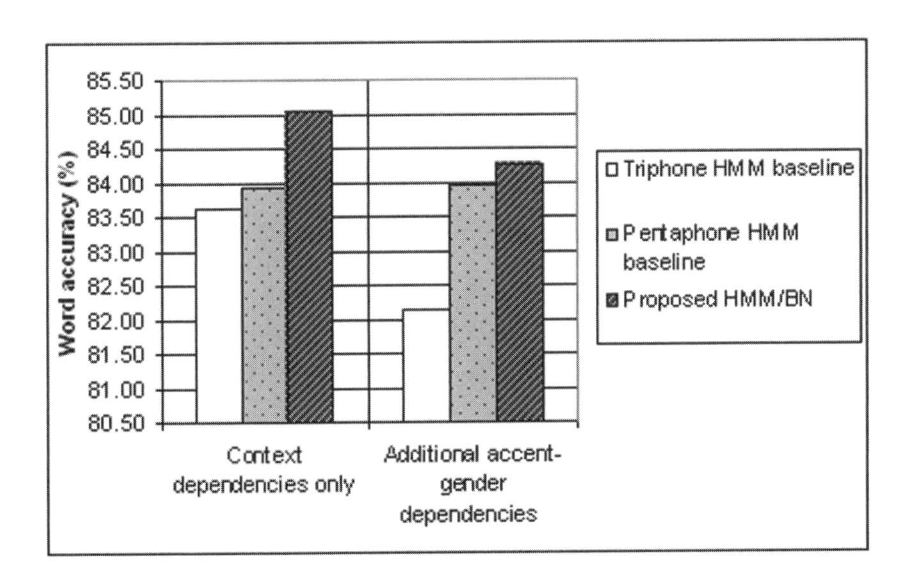

Fig. 4.24. Comparing recognition accuracy rates of different systems: triphone HMM baseline, pentaphone HMM baseline, and the proposed pentaphone HMM/BN models having the same five mixture components per state.

In the case of context dependencies only, the pentaphone HMM baseline produced slight improvement over the triphone HMM baseline. This may be due to the sparseness of training data. The resolution of the context was then reduced, since there seemed to be too many different pentaphone contexts sharing the same Gaussian components. By incorporating the pentaphone knowledge in the triphone state PDF by means of a BN, a better performance could be obtained. The performance of the pentaphone HMM baseline did not decrease when gender and accent were incorporated, as was the case for the triphone HMM baseline and the proposed HMM/BN model, which is probably due to the use of interpolation in the rescoring process. However, the best performance was still obtained by the proposed HMM/BN model.

4.3.2 Incorporating Knowledge at the HMM Phonetic-unit Level

We have applied GFIKS at the HMM phonetic-unit level following the common consideration described in Section 4.1. Here, we have also attempted to incorporate various additional knowledge sources, including wide-phonetic context, accent and gender information.

Incorporating Wide-phonetic Context Information

The GFIKS has been applied to the problem of extending triphone context $/a^{-}, a, a^{+}/$ to the pentaphone $/a^{--}, a^{-}, a, a^{+}, a^{++}/$.

a. Causal Relationship and Inference

We incorporate the additional second preceding context C_L of $/a^{--}/$ and succeeding context C_R of $/a^{++}/$ into the probability function $P(X_s|\lambda)$ using the proposed GFIKS framework. The causal relationship among X_s, λ, C_L, and C_R is described by the BN shown in Figure 4.25 (similar to the one in Figure 3.13(a)).

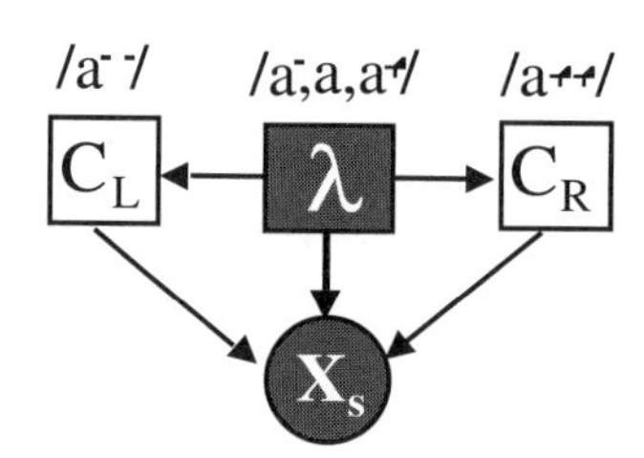

Fig. 4.25. BN topology structure describing the conditional relationship among X_s, λ, C_L, and C_R.

Here, we decompose $P(X_s|C_L, C_R, \lambda)$ using the junction tree algorithm as described in Section 3.2.3, since it is difficult to obtain a simple functional form for this conditional PDF. Figure 4.26 shows the equivalent BN topology, the moral and triangulated graph and also the final junction tree, which is similar to the one in Figure 3.13, where M is our current HMM phonetic model, λ, and D is segment X_s.

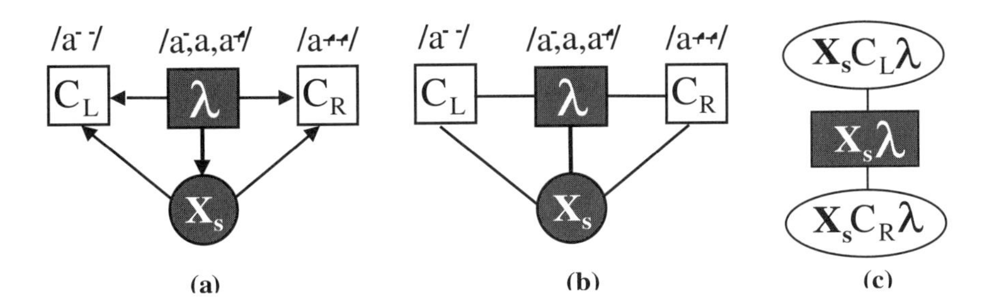

Fig. 4.26. (a) Equivalent BN topology. (b) Moral and triangulated graph of Figure 4.26(a). (c) Junction tree of Figure 4.26(b).

The conditional probability function is then defined as

$$P(X_s|C_L, C_R, \lambda) = \frac{P(X_s|C_L, \lambda)P(X_s|C_R, \lambda)}{P(X_s|\lambda)}, \tag{4.24}$$

according to Eq. (4.6). Since λ is associated with the triphone $/a^-, a, a^+/$, the second preceding C_L with $/a^{--}/$, and the second succeeding C_R with $/a^{++}/$, we can write

$$\begin{aligned} &P(X_s|C_L, C_R, \lambda) \\ &= P(X_s|a^{--}, a^{++}, [a^-, a, a^+]) \\ &= P(X_s|[a^{--}, a^-, a, a^+, a^{++}]), \end{aligned} \tag{4.25}$$

and Eq. (4.25) becomes

$$\begin{aligned} &P(X_s|[a^{--}, a^-, a, a^+, a^{++}]) \\ &= \frac{P(X_s|a^{--}, [a^-, a, a^+])P(X_s|a^{++}, [a^-, a, a^+])}{P(X_s|[a^-, a, a^+])} \\ &= \frac{P(X_s|[a^{--}, a^-, a, a^+])P(X_s|[a^-, a, a^+, a^{++}])}{P(X_s|[a^-, a, a^+])}. \end{aligned} \tag{4.26}$$

This indicates that a pentaphone model $P(X_s|[a^{--}, a^-, a, a^+, a^{++}])$ may be composed of the less complex models, including $p(X_s|[a^{--}, a^-, a, a^+])$, $p(X_s|[a^-, a, a^+, a^{++}])$, and $p(X_s|[a^-, a, a^+])$. They correspond to the likelihood of segment X_s given the left/preceding-tetraphone-context (L4), right/ following-tetraphone-context (R4), and center-triphone-context units (C3), the so-called C3L4R4 composition. However, developing tetraphone models for $[a^{--}, a^-, a, a^+]$ and $[a^-, a, a^+, a^{++}]$ may also be difficult due to the sparsity of data.

Instead, let us use Eq. (4.25) and adjust λ to represent a monophone $/a/$, and the second preceding and succeeding contexts, C_L and C_R, to respectively represent $/a^{--}, a^-/$ and $/a^+, a^{++}/$. Then,

$$\begin{aligned} &P(X_s|[a^{--}, a^-, a, a^+, a^{++}]) \\ &= \frac{P(X_s|[a^{--}, a^-], a)P(X_s|[a^+, a^{++}], a)}{P(X_s|[a])} \\ &= \frac{P(X_s|[a^{--}, a^-, a])P(X_s|[a, a^+, a^{++}])}{P(X_s|[a])}, \end{aligned} \tag{4.27}$$

which indicates that the pentaphone-context, $/a^{--}, a^-, a, a^+, a^{++}/$, is composed of $p(X_s|[a^{--}, a^-, a])$, $p(X_s|[a, a^+, a^{++}])$, and $p(X_s|[a])$, which correspond to the likelihood of observation X_s given the left/preceding-triphone-context unit (L3), the right/following-triphone-context unit (R3), and the monophone unit (C1). We call this composition C1L3R3, and it is shown structurally in Figure 4.27(c).

As can be seen, the number of context units to be estimated is reduced from N^5 to $(2N^3 + N)$, without loss of context coverage, where N is the number of phones. If we use a 44-phoneme set for English ASR, the total number of different contexts that need to be estimated in the pentaphone model is $44^5 =\sim 165,000,000$ context units. A composition with triphone-context units reduces the complexity to about 170,000 context units.

Structurally, the conventional HMM of a triphone-context unit model can be described as in Figure 4.27(a), and that of a pentaphone-context unit model can be described as in Figure 4.27(b).Analyzing Eqs. (4.26) and (4.27), we can see that Eq. (4.24) can be used as a starting point for deriving other compositions of the HMM phonetic model as well. When we assume that λ is a monophone unit $/a/$ and that C_L and C_R are preceding and following context units $/a^-/$ and $/a^+/$, respectively, we can obtain the same factorization as that proposed by Ming et al. (1999); Ming and Smith (1998) previously, known as the Bayesian triphone:

$$P(X_s|[a^-, a, a^+]) = \frac{P(X_s|[a^-, a])P(X_s|[a, a^+])}{P(X_s|[a])}, \tag{4.28}$$

where the triphone model is constructed from monophone and biphone models. Hereinafter, any model composed in this way will also be called a Bayesian models.

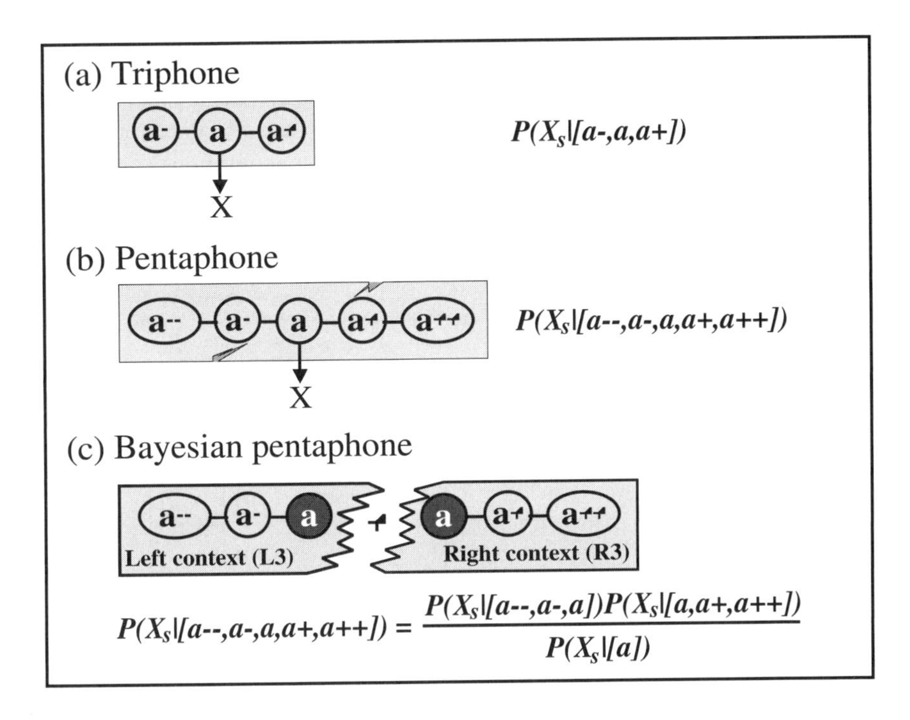

Fig. 4.27. (a) Conventional triphone model, (b) Conventional pentaphone model, (c) Bayesian pentaphone model composition C1L3R3, consisting of the preceding/following triphone-context unit and center-monophone unit.

The extended version of the Bayesian triphone, the so-called Bayesian wide-phonetic context model, can also be found in our previous study (Sakti et al., 2006, 2005). This approach allows us to model a wide range of phonetic contexts from less context-dependent models simply based on Bayes's rule (See Appendix C for more details). However, difficulties arise when different types of knowledge sources need to be incorporated.

b. Enhancing Model Reliability

To enhance model reliability, we apply the three smoothing techniques described in Section 4.2.3, including the "No decision," "hard decision," and "soft decision" mechanisms using deleted interpolation.

c. Training and Recognition Issues

All components of the pentaphone $P(X_s|[a^{--}, a^{-}, a, a^{+}, a^{++}])$ model have been trained separately, given the segment observation data X_s and the respective specific additional phonetic context.

The proposed pentaphone models are applied by rescoring the N-best list generated by a standard triphone-based HMM system as described in Section 4.2.5. A block diagram of rescoring using the pentaphone composition models is given in Figure 4.28.

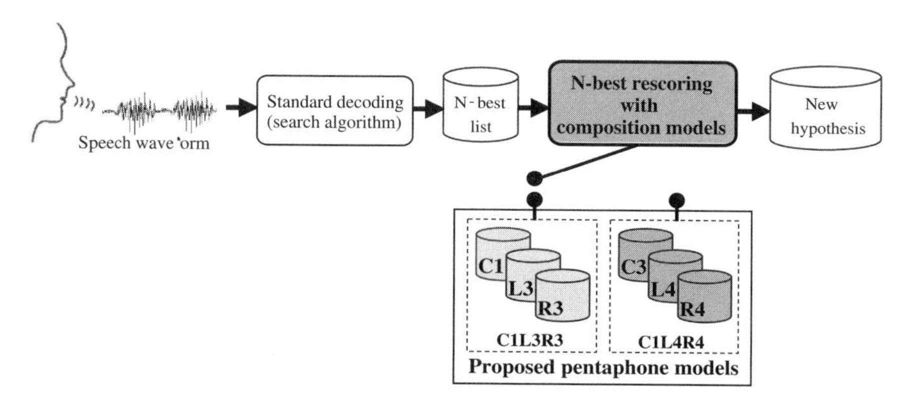

Fig. 4.28. Rescoring procedure with pentaphone composition models: C1L3R3 or C3L4R4.

Figure 4.29 illustrates the N-best rescoring mechanism, where every phoneme segment in each hypothesis is rescored using the proposed pentaphone models. At the beginning/end of the utterance, all left/right contexts are filled by silence. Since we assume that there is no long silence between adjacent words, the final phonetic context of the previous word will also affect the beginning phonetic context of the current word. Thus, this rescoring mechanism is performed in the same way for each segment within and between words (cross-word model). Then, the new scores are combined with the LM score for this hypothesis. The hypothesis achieving the highest total utterance score among the N-best is selected as the new recognition output.

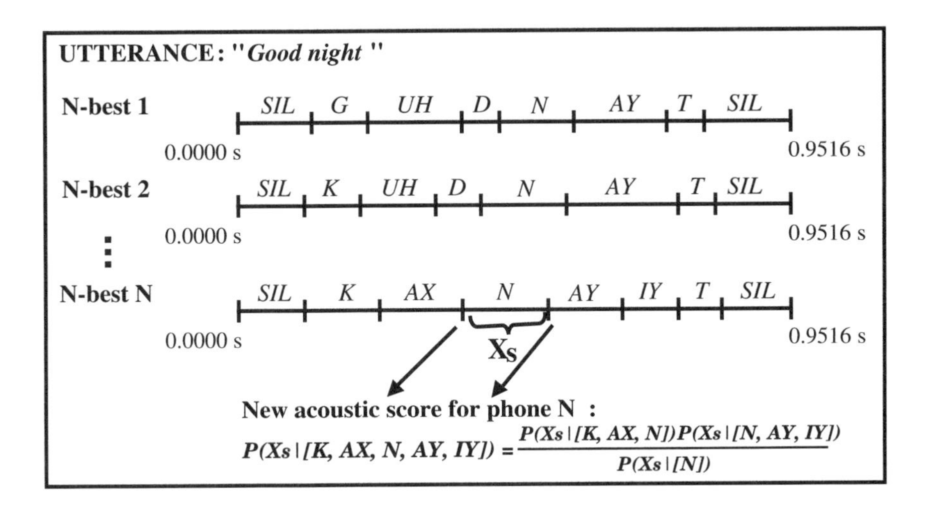

Fig. 4.29. N-best rescoring mechanism.

d. Experimental Set-Up

Similar to the experiments described in Section 4.3.1, we have conducted experiments on the specific ATR BTEC task using the ATR basic parameter set-up.

e. Accuracy of Recognition

As training data, we again used 60 hours of native English speech data from the SI-284 Wall Street Journal (WSJ0 and WSJ1) speech corpus (Paul and Baker, 1992). Each component model of the Bayesian wide-phonetic context models is trained separately using the same SSS training algorithm, the same amount of training data, and the same number of 15 Gaussian mixture components per state as our optimum choice.

The performances of the models have been evaluated on the ATR BTEC task (Takezawa et al., 2002), which is quite different from the training corpus. The full BTEC test set1 consists of 4,080 read speech utterances spoken by 40 different speakers (20 Males, 20 Females). In this study, in order to reduce the training time, we simply selected 1,000 utterances spoken by 20 different speakers (10 Males, 10 Females), and used them as a development set to find the optimum λ parameter of the deleted interpolation. Two hundred randomly selected utterances spoken by 40 different speakers (20 Males, 20 Females) were used as a test set.

In the first experiment, a context-independent monophone system with 132 total states was used as a baseline. Rescoring was done with the Bayesian triphone C1L2R2 (with 2,700 states, sum of C1: 132 st., L2: 1,313 st., and R2: 1,255 st.) as described in Section 4.2.5. For comparison, we also rescored using the conventional biphone C2 (1,313 states) and triphone C3 (2,009 states) models. In each rescoring, we applied "no decision," "hard decision," and "soft decision" mechanisms (see Section 4.2.3).

The recognition results for all models, obtained with each decision mechanism, are shown in Figure 4.30. Bayesian C1L2R2 could achieve a significant improvement of 5.6% relative to the baseline. Its performance is better than the biphone C2 alone but still worse than the triphone C3. According to our preliminary experiments, the WSJ database is more or less suitable for training a conventional triphone model without losing the context resolution, and the optimum model was the one with about 2,000-3,000 total states and 15 Gaussian mixture components. Therefore, C3, with 2,009 states and 15 mixture components per state, is optimum in terms of parameter number and context resolution. This may be the reason why it yields the best result.

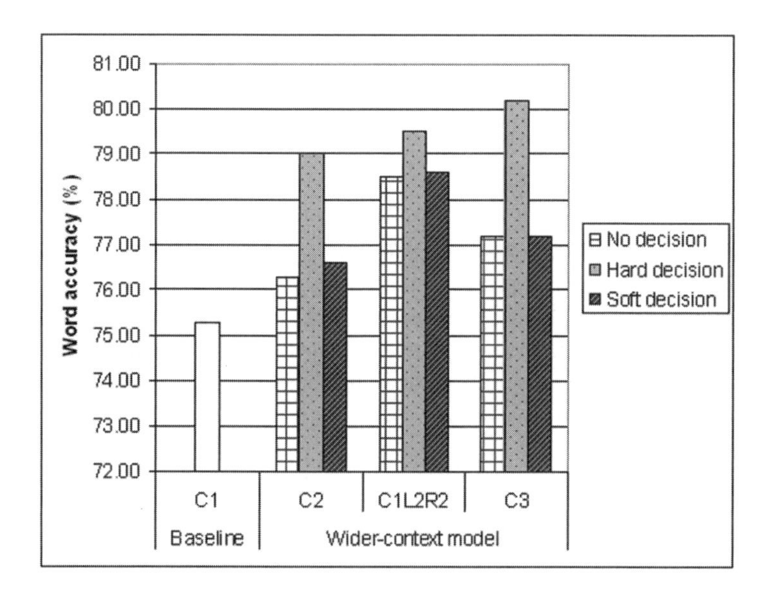

Fig. 4.30. Recognition accuracy rates of Bayesian triphone model.

The best performances have been obtained using the "hard decision" mechanism. This resulted in a higher performance than using the "no decision" and "soft decision" mechanisms with the optimal weight parameter $\lambda = 0.5$. This may be due to the following reasons. Considering the amount of training data and the number of parameters, the triphone model is optimal and much more precise than the monophone model. But it may give an unreliable estimation if there are some unseen phonetic contexts in the testing data. Accordingly, the "hard decision," or back-off smoothing, seems to be the optimal choice because it only falls back to the monophone model if the output from the triphone model is unreliable. On the other hand, using the "no decision" mechanism, which always accepts the output value from the triphone model, may contain some unreliable outputs due to unseen contexts, and using the "soft decision" mechanism, which always interpolates the triphone model with the monophone model with equal weight ($\lambda = 0.5$), may hurt the recognition accuracy of the triphone model.

Next, we experimented with wider context models, where we used the context-dependent triphone system with 2,009 total states as the baseline to generate new N-best lists for rescoring. As described in part (a.), "Causal Relationship and Inference," there are two types of Bayesian pentaphone models: composition C1L3R3 and composition C3L4R4. We also included two other Bayesian pentaphone models: C1Lsk3Rsk3 and C1C3Csk3 (the details of these compositions can be found in Appendix C.2), as well as a conventional full pentaphone model (C5) trained from scratch. The C1L3R3 model has 3,175 states (sum of C1: 132 st., L3: 1,524 st., R3: 1,519 st.), the C3L4R4 model has 6,052 states (sum of C3: 2,009 st., L4: 2,021 st., R4: 2,022 st.), the C1Lsk3Rsk3 model has 3,333 states (sum of C1: 132 st., Lsk3: 1,587 st., Rsk3: 1,614 st.), the C1C3Csk3 model has 3,250 states (sum of C1: 132 st., C3: 2,009 st., Csk3: 1,109 st.), and, finally, the full pentaphone C5 model has 2,040 total states. The recognition results for all models, obtained by each decision mechanism, are shown in Figure 4.31.

All of the pentaphone models could also achieve improvement relative to the baseline. Here, the conventional pentaphone C5 gives a worse performance than the Bayesian pentaphone models. This may be due to the following reason. Given the amount of the WSJ training data, the optimum pentaphone model achieved with the MDL-SSS algorithm has 2,040 total states, which is not so different from the total number of states in the triphone C3. It seems that there are many different pentaphone contexts sharing the same Gaussian components, and thus the context resolution is reduced. Accordingly, approximating a pentaphone model using the Bayesian composition of several less-context-dependent models such as triphone models could help to reduce the loss of context resolution and improve performance.

Among the Bayesian pentaphone models, the C1L3R3 model gives the best result, and the worst is from the C3L4R4 model. The reason for this may be that the WSJ training data are also not sufficient to properly train the model parameters of the tetraphone components L4 and R4. Their total number of states is only slightly different from the total number of states of triphone C3. Consequently, as has been the case for C5, there may be many tetraphone contexts that share the same Gaussian components, thus reducing the context resolution. Another reason may be that the triple phoneme overlap between the L4 and R4 component models is too large, so composing these models could not give an optimum solution. However, the other Bayesian models, C1L3R3, C1Lsk3Rsk3, and C1C3Csk3, yield a similar number of total states, and the total amount of training data would be sufficient to train the triphone contexts.

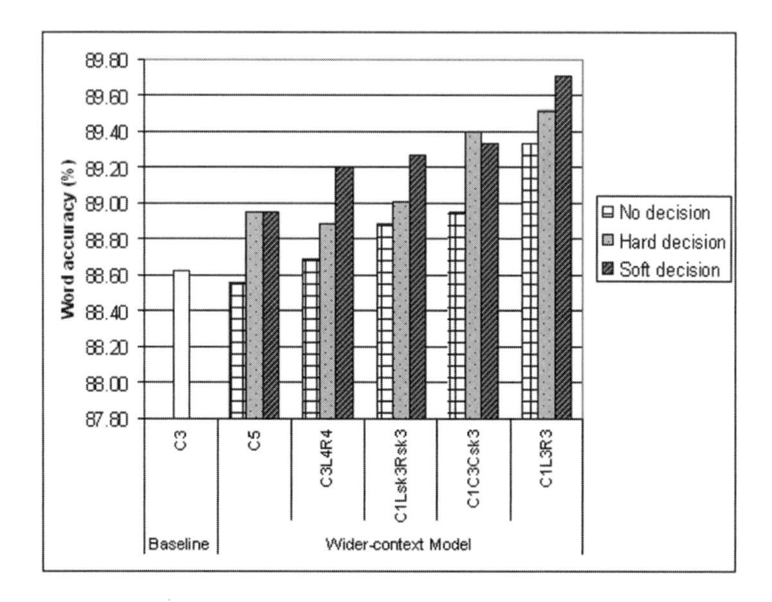

Fig. 4.31. Recognition accuracy rates of Bayesian pentaphone models.

Moreover, the C1L3R3, C1Lsk3Rsk3, and C1C3Csk3 compositions have only a single-phoneme overlap between the model components. However, considering a context's phonetic dependency, the dependency between adjacent phonetic contexts may have much stronger effects than the dependency between skipped phonetic contexts. This means that the more adjacent phonetic contexts a model has, the better the model. Thus, C1C3Csk3 is better than C1Lsk3Rsk3, and the C1L3R3 model is the best among all of the models. The differences are significant at the 5% level calculated using the Sign test (see Appendix D).

In this case, the best performance has been obtained by the "soft decision" mechanism using deleted interpolation. This shows that, if the estimation of the pentaphone model is less reliable, it is useful to interpolate the pentaphone model and the triphone model estimations, since the triphone model can often provide useful information. The optimal weight parameter λ has been about 0.3. Having a weight factor of 0.3 means that the contribution of the pentaphone model is only about 30% of the total score. However, even with this relatively small contribution, the results show that it can still help to improve recognition performance.

Figure 4.32 compares the relative reduction in WER achieved by the Bayesian triphone C1L2R2 model from the monophone baseline with that by the Bayesian pentaphone C1L3R3 model from the triphone baseline. The reduction in error rate by the Bayesian pentaphone model is smaller, at about half the reduction in error rate by the Bayesian triphone model, probably due to the following reasons. First, the coarticulation effect from the second preceding and following contexts is less than the coarticulation effect from the first preceding and following contexts. Second, the variations in the read speech data due to longer coarticulation effects may be less than in conversational speech. This can also be seen from the weight factor of the deleted interpolation, which can be interpreted as a confidence factor of only 30%. However, even with this relatively small contribution, the results show that it can still help to improve recognition performance.

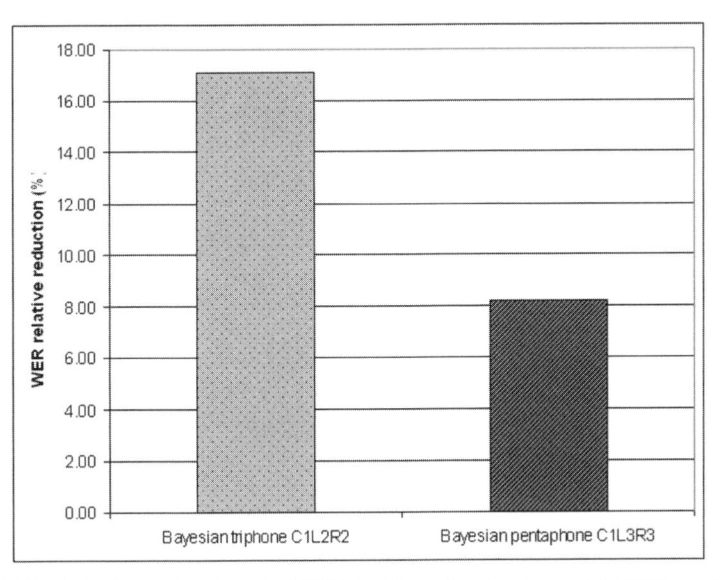

Fig. 4.32. Relative reductions in WER by Bayesian triphone C1L2R2 model from monophone baseline and by Bayesian pentaphone C1L3R3 model from triphone baseline.

To show the consistency of the effect of using the Bayesian composition, we have performed another experimental evaluation on fewer training data. Here, we chose the TIMIT acoustic-phonetic continuous speech corpus (Garofolo et al., 1993) as another American-English and phonetically rich corpus, but one smaller than the WSJ database corpus (see the brief information in Appendix A.2). It contains only about seven hours of read speech (6,300 utterances in total). Each component acoustic model has been trained using the SSS algorithm as before. In this case, the triphone baseline has 434 states, the conventional pentaphone C5 has 440 states, and the proposed Bayesian pentaphone C1L3R3 has 850 states (sum of C1: 132 st., L3: 369 st., R3: 349 st.). These models have been tested using the same BTEC test set with "soft decision" only. The optimal weight parameter λ was also 0.3. The results are shown in Figure 4.33. As can be seen, with fewer training data, the performance difference between the proposed C1L3R3 model and the conventional pentaphone C5 model became more significant.

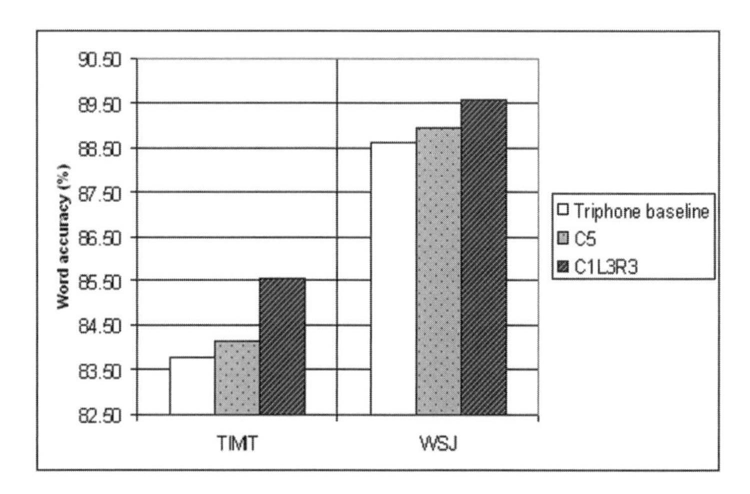

Fig. 4.33. Recognition accuracy rates of conventional pentaphone C5 and proposed Bayesian pentaphone C1L3R3 models with different amounts of training data.

Incorporating Multiple Knowledge: Accent, Gender and Wide-phonetic Context Information

Our proposed GFIKS unified framework gives us a more appropriate means of incorporating various kinds of knowledge sources. For example, we can easily further extend C1L3R3 with other additional knowledge variables, such as gender or accent information. We can extend C1L3R3 with gender information only (C1L3R3-G), with accent information only (C1L3R3-A), or with both (C1L3R3-AG).

a. Causal Relationship and Inference

For the case of C1L3R3-AG, the causal relationship among X_s, λ, C_L, C_R, A and G is described by the BN shown in Figure 4.34.

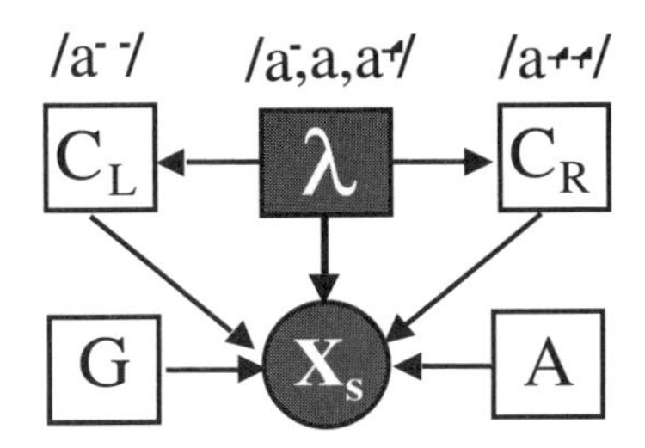

Fig. 4.34. BN topology structure describing the conditional relationship among X_s, λ, C_L, C_R, A, and G.

Here, we decompose $P(X_s|C_L, C_R, \lambda, A, G)$ using the junction tree algorithm described in Section 3.2.3, where the equivalent BN topology, moral and triangulated graph, and corresponding junction tree are shown in Figure 4.35.

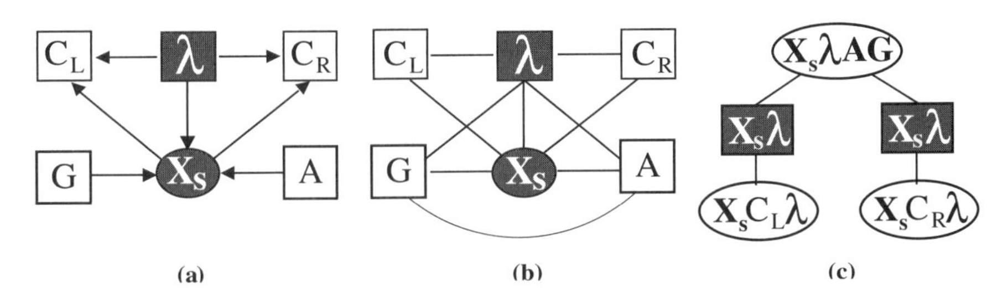

Fig. 4.35. (a) Equivalent BN topology of Figure 4.34. (b) Moral and triangulated graph of Figure 4.35(a). (c) Corresponding junction tree.

The conditional probability function is obtained as

$$\begin{aligned}
&P(X_s|C_L, C_R, \lambda, A, G) \\
&= P(X_s|\lambda, A, G)\frac{P(X_s|C_L, \lambda)}{P(X_s|\lambda)}\frac{P(X_s|C_R, \lambda)}{P(X_s|\lambda)} \\
&= \frac{P(X_s|\lambda, A, G)P(X_s|C_L, \lambda)}{P(X_s|\lambda)}\frac{P(X_s|\lambda, A, G)P(X_s|C_R, \lambda)}{P(X_s|\lambda)} \\
&\quad \cdot \frac{1}{P(X_s|\lambda, A, G)} \\
&= \frac{P(X_s|C_L, \lambda, A, G)P(X_s|C_R\lambda, A, G)}{P(X_s|\lambda, A, G)}.
\end{aligned} \tag{4.29}$$

Accordingly, following the same setting as C1L3R3 for λ, C_L, and C_R, the pentaphone likelihood of C1L3R3-AG becomes

$$P(X_s|[a^{--}, a^{-}, a, a^{+}, a^{++}], A, G) = \frac{P(X_s|[a^{--}, a^{-}, a], A, G)P(X_s|[a, a^{+}, a^{++}], A, G)}{P(X_s|[a], A, G)}, \quad (4.30)$$

which indicates that $P(X_s|[a^{--}, a^{-}, a, a^{+}, a^{++}], A, G)$ can be simplified by factorizing it into $P(X_s|[a^{--}, a^{-}, a], A, G)$, $P(X_s|[a, a^{+}, a^{++}], A, G)$, and also $P(X_s|[a], A, G)$.

b. Enhancing Model Reliability

To enhance the model reliability, we have applied the three smoothing techniques described in Section 4.2.3, including the "No decision," "hard decision," and "soft decision" mechanisms using deleted interpolation.

c. Training and Recognition Issues

All components of the accent-gender-dependent pentaphone model have been trained separately given the segment observation data X_s and the respective additional phonetic context, accent and gender information. The proposed accent-gender-dependent pentaphone models have been applied by rescoring the N-best list generated by a standard triphone-based HMM system as described in Section 4.2.5. A block diagram of rescoring using accent-gender-dependent pentaphone composition models is given in Figure 4.36.

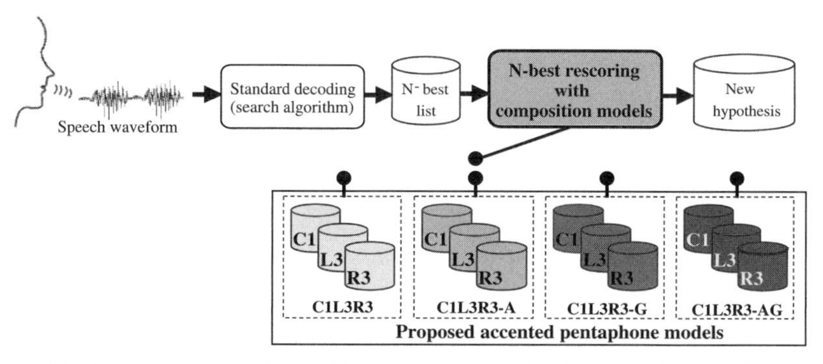

Fig. 4.36. Rescoring procedure with the accent-gender-dependent pentaphone composition models: C1L3R3, C1L3R3-A, C1L3R3-G, and C1L3R3-AG.

d. Experimental Set-Up

Similar to the experiments described in Section 4.3.1, we have conducted an experiment on the accented speech of the ATR BTEC task using the

ATR basic parameter set-up.

e. Accuracy of Recognition

We have investigated several different ways of decomposing pentaphone models and found that the best was C1L3R3 composition (Sakti et al., 2006, 2005). Here, we describe additional experiments using only the C1L3R3 model.

All components of all accented pentaphone models have been trained separately using the same amount of training data and the same SSS training algorithm. There have been a total of 3,660 states (sum of C1: 132 states, L3: 1,746 states, R3: 1,782 states) with four different numbers of Gaussian mixture components per state, i.e., 5, 10, 15, and 20. An embedded training procedure was then carried out for pentaphones C1L3R3-A, C1L3R3-G, and C1L3R3-AG with specific accent or gender training data.

The performance for each model with 5, 10, or 20 mixture components per state is shown in Figure 4.37. For the case of 5 mixture components per state, the triphone baseline without any additional knowledge achieved 83.60% word accuracy. However, for the accent-dependent models, word accuracy decreased to 82.11%. This may be due to the size of training data, which is much smaller compared to the other baseline models. By rescoring with a more precise pentaphone model we would expect the performance to improve. The differences are significant at the 5% level calculated using the Sign test (see Appendix D). There is no performance decrease when gender and accent are incorporated, as in the case of the triphone baseline, which is probably due to the use of deleted interpolation.

Among the pentaphone models, the performance of the proposed model has always been better than that of the conventional pentaphone HMM. This may be because, given the amount of training data, the training of the conventional pentaphone model using the MDL-SSS algorithm resulted in a model having 2,202 total states, which is not so different from the total number of states in the triphone HMM. It seems that there have been many different pentaphone contexts sharing the same Gaussian components, and thus the context resolution has been reduced. Consequently, approximating a pentaphone model using the composition of several less-context-dependent models could help to reduce the loss of context resolution and improve performance. The best performance has been obtained by the model that incorporated additional knowledge of accent A, gender G, second preceding context C_L, and succeeding context C_R.

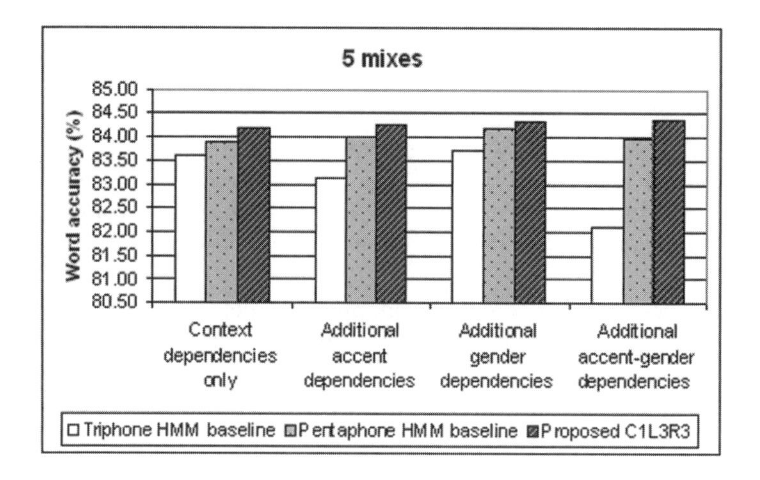

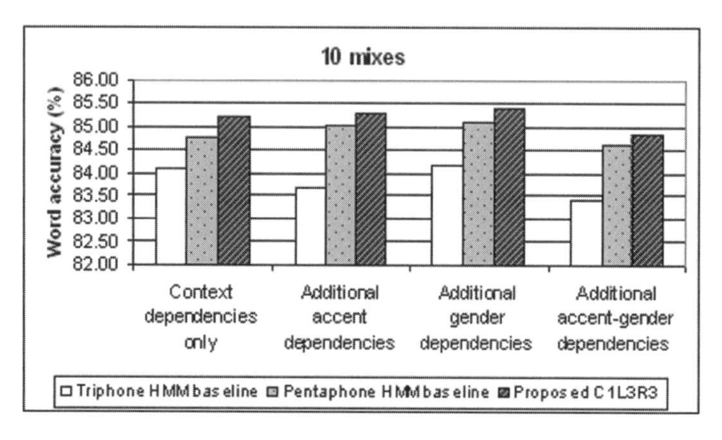

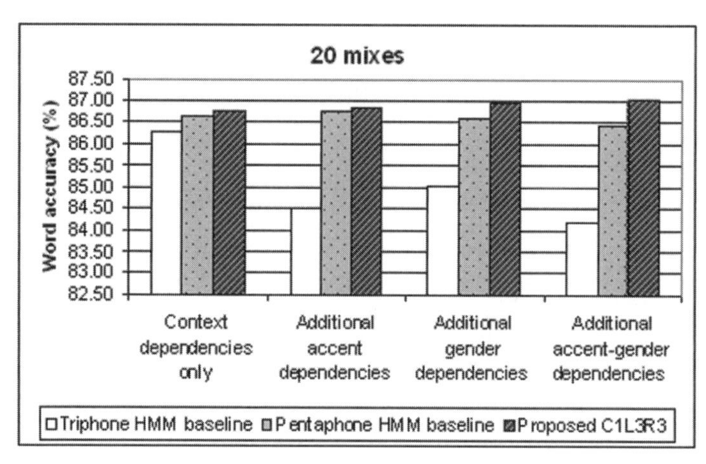

Fig. 4.37. Comparing recognition accuracy rates of different systems triphone HMM baseline, pentaphone HMM baseline, and proposed pentaphone models having the same 5, 10, and 20 mixture components per state.

Overall, through using different mixture components per state, the results show that the proposed pentaphone models consistently outperform the standard HMM baseline.

We then investigated in more detail how well C1L3R3-AG performed on all accented test data, using the N-best (N=10) list. The weight parameter for deleted interpolation, λ, was the same (0.3). Here, we measured both the relative improvement (Rel-Imp) and the relative improvement to rescoring (Rel-Resc-Imp) as used by Li et al. (2005):

$$RelRescImp = \frac{\text{Rescoring Result} - \text{Baseline}}{\text{N-best List Upper Bound} - \text{Baseline}}, \tag{4.31}$$

where the N-best list upper bound is the N-best recognition result.

The results obtained with models having different numbers of mixture components are summarized in Table 4.8. As can be seen, the proposed pentaphone model consistently improved the performance of the ASR system. The largest Rel-Resc-Imp has been achieved with 15 mixture models for both US- and AUS-accented models at 37.92% and 38.04%, respectively.

Table 4.8. *Recognition accuracy rates (%) for proposed Bayesian pentaphone C1L3R3-AG (see Eq. (4.30)) on a test set of matching accents with different numbers of mixture components.*

	US accent				
Mixture number	**Upper bound**	**Triphn baseline**	**Proposed C1L3R3-AG**	*Rel Imp*	*Rel Resc-Imp*
5 mix	87.52	84.30	85.19	*5.67*	*27.64*
10 mix	87.94	84.66	85.79	*7.37*	*34.45*
15 mix	87.76	84.78	85.91	*7.42*	*37.92*
20 mix	87.78	85.25	85.91	*4.47*	*26.09*

	AUS accent				
Mixture number	**Upper bound**	**Triphn baseline**	**Proposed C1L3R3-AG**	*Rel Imp*	*Rel Resc-Imp*
5 mix	85.79	82.33	83.76	*8.09*	*41.33*
10 mix	85.37	82.21	82.81	*3.37*	*18.99*
15 mix	86.93	83.46	84.78	*7.98*	*38.04*
20 mix	86.39	82.63	83.58	*5.47*	*25.27*

We have also evaluated how the proposed pentaphone C1L3R3-AG model performs on a test set of mismatched accents. The results obtained using a model with 15 mixture components are summarized in Table 4.9.

The results from evaluating matching accents are also included to allow easy comparison. We can see that, the proposed pentaphone C1L3R3-AG model also consistently outperforms the standard triphone model with mismatched accents.

Table 4.9. *Recognition accuracy rates (%) for proposed Bayesian pentaphone C1L3R3-AG model (see Eq. (4.30)) on a test set of mismatched accents with 15 mixture components.*

	US accent				
Accented test set	**Upper bound**	**Triphn baseline**	**Proposed C1L3R3-AG**	*Rel Imp*	*Rel Resc-Imp*
US test	87.76	84.78	85.91	*7.42*	*37.92*
AUS test	71.76	64.78	68.12	*9.48*	*47.85*

	AUS accent				
Accented test set	**Upper bound**	**Triphn baseline**	**Proposed C1L3R3-AG**	*Rel Imp*	*Rel Resc-Imp*
US test	80.60	75.22	77.31	*8.43*	*38.85*
AUS test	86.93	83.46	84.78	*7.98*	*38.04*

4.4 Experiments Summary and Discussion

In this chapter, we have demonstrated the incorporation of various knowledge sources at the HMM state level. This method allows for easy integration of additional information into existing HMM-based triphone acoustic models, where additional knowledge sources are incorporated in the triphone state PDF by means of the BN. In this approach, the temporal speech characteristics are still governed by HMM state transitions, but HMM state probability distributions are inferred from a BN, which is why this is called the HMM/BN model. For issues of recognition, if we lack appropriate decoding for the pentaphone HMM/BN models, we can still use the standard decoding system without modification, in which case the additional knowledge sources are assumed hidden, and the state PDF can be calculated by marginalization over those BN joint PDFs.

The recognition results reveal that ASR system performance can be improved by incorporating wide-phonetic context units based on our proposed framework in the term of a hybrid pentaphone HMM/BN model. This shows that, by changing only the probability distribution, we can obtain improved recognition, even when the model has the same number of Gaussians as the baseline triphone HMM. The best performance among the pentaphone HMM/BN models has been obtained by the one that incorporated additional knowledge of second preceding context C_L and succeeding context C_R.

We have also demonstrated the incorporation of various knowledge sources at the HMM phonetic-unit level by constructing a model from several less complex models. The composition framework is based on a junction tree algorithm, which leads to a reduction in the number of context units to be estimated, and thus a significant reduction in the loss of context resolution. We applied these composition models at the post-processing stage with N-best rescoring. Experimental results reveal that our method improves word accuracy compared to the standard HMM with or without additional knowledge sources. The best performance has been obtained by the model that incorporates additional knowledge of accent A, gender G, second preceding context C_L, and second succeeding context C_R.

Summaries of the experiments on incorporating various knowledge sources at both the HMM state and phonetic-unit level can be seen in Table 4.10 and Table 4.11, respectively. These tables' summarization includes the following descriptions:

1. Type of knowledge sources

   This includes background noise, gender, accent and wide-phonetic context information.

2. BN topology

   This defines the causal relationship between information sources.

3. Inference

   This may be done directly on BN or using junction tree decomposition

4. ASR task and performance

   - Type of ASR task

     This includes small vocabulary and LVCSR task.

   - Type of speech data

     This includes clean speech, noisy speech and accented speech.

   - Training data including total hours of speech.

   - Test data.

   - Recognition performance and significance level.

Table 4.10. *Summary of incorporating various knowledge sources at the HMM state level.*

Knowledge sources	**BN topology**	**Inference**	**ASR task and performance**
Gender (G)	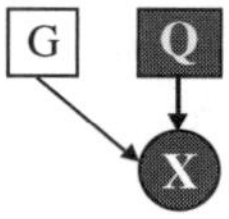 $P(X,G,Q) =$ $P(X\|G,Q)P(G)P(Q)$	Direct inference (G is assumed hidden) $p(x_t\|q_j) =$ $\sum_{m=1}^{M} p(g_m)p(x_t\|g_m,q_j)$	• LVCSR (official ARPA benchmark test) • Clean speech • WSJ SI-284 ($\sim$ 60 hours of speech) • WSJ Hub2-5k • Performed the 3rd world rank
Noise (N), SNR (S)	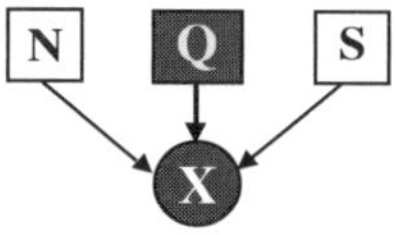 $P(X,N,S,Q) =$ $P(X\|N,S,Q)P(N)\cdot$ $P(S)P(Q)$	Direct inference (N & S are assumed hidden) $p(x_t\|q_j) =$ $\sum_{m=1}^{M_N}\sum_{n=1}^{M_S} p(n_m)p(s_n)\cdot$ $p(x_t\|n_m,s_n,q_j)$	• Connected digit (official AURORA task) • Noisy speech • AURORA2 ($\sim$ 4 hours of speech) • AURORA2 Set A&B • Performed the best (Better than HMM & DBN, specially in low SNR)
Wide-phonetic context (C_L, C_R)	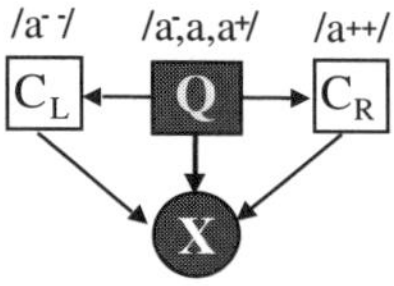 $P(X,C_L,C_R,Q) =$ $P(X\|C_L,C_R,Q)\cdot$ $P(C_L\|Q)P(C_R\|Q)P(Q)$	Direct inference (C_L & C_R are assumed hidden) $p(x_t\|q_j) =$ $\sum_{l=1}^{M_L}\sum_{r=1}^{M_R} p(c_l\|q_j)p(c_r\|q_j)\cdot$ $p(x_t\|c_l,c_r,q_j)$	• LVCSR • Clean speech • WSJ SI-284 ($\sim$ 60 hours of speech) • ATR BTEC (mismatched test set) • Up to 10% relative WER reduction • Significant (Sign test $\alpha = 0.05$) $p \leq 0.00226$
Accent (A), gender (G), wide-phonetic context (C_L, C_R)	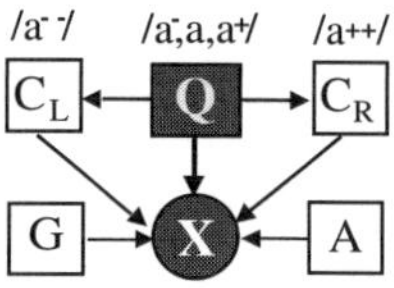 $P(X,C_L,C_R,Q,A,G) =$ $P(X\|C_L,C_R,Q,A,G)\cdot$ $P(C_L\|Q)P(C_R\|Q)\cdot$ $P(Q)P(A)P(G)$	Direct inference (C_L, C_R, A, & G are assumed hidden) $p(x_t\|q_j) =$ $\sum_{n=1}^{M_A}\sum_{m=1}^{M_G}\sum_{l=1}^{M_L}\sum_{r=1}^{M_R} p(a_n)p(g_m)\cdot$ $p(c_l\|q_j)p(c_r\|q_j)\cdot$ $p(x_t\|c_l,c_r,q_j,a_n,g_m)$	• LVCSR • Clean accented speech • ATR accented EDB ($\sim$ 80 hours of speech) • ATR accented EDB • Up to 12% relative WER reduction • Significant (Sign test $\alpha = 0.05$) $p \leq 0.00289$

Table 4.11. *Summary of incorporating various knowledge sources at the HMM phonetic unit level.*

Knowledge sources	BN topology	Inference	ASR task and performance
Wide-phonetic context (C_L,C_R)	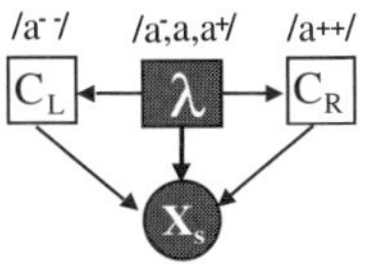 /a⁻/ /a⁻,a,a⁺/ /a⁺⁺/ C_L λ C_R X_s $P(X_s, C_L, C_R, \lambda) = P(X_s \| C_L, C_R, \lambda) \cdot P(C_L \| \lambda) P(C_R \| \lambda) P(\lambda)$	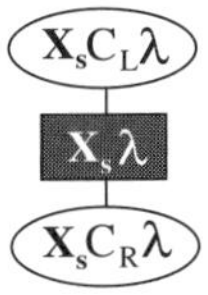 $X_sC_L\lambda$ $X_s\lambda$ $X_sC_R\lambda$ $P(X_s \| C_L, C_R, \lambda) = \frac{P(X_s \| C_L, \lambda) P(X_s \| C_R \lambda)}{P(X_s \| \lambda)}$ (Using junction tree*)	• LVCSR • Clean speech • WSJ SI-284 (~ 60 hours of speech) • ATR BTEC (mismatched test set) • Up to 9.5% relative WER reduction • Significant (Sign test $\alpha = 0.05$) $p \leq 0.000488$
Accent (A), gender (G), wide-phonetic context (C_L,C_R)	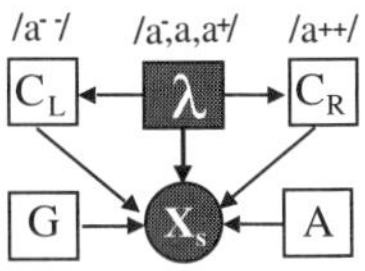 /a⁻/ /a⁻,a,a⁺/ /a⁺⁺/ C_L λ C_R G X_s A $P(X_s, C_L, C_R, \lambda, A, G) = P(X_s \| C_L, C_R, \lambda, A, G) \cdot P(C_L \| \lambda) P(C_R \| \lambda) \cdot P(\lambda) P(A) P(G)$	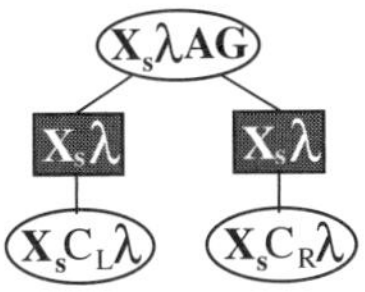 $X_s\lambda AG$ $X_s\lambda$ $X_s\lambda$ $X_sC_L\lambda$ $X_sC_R\lambda$ $P(X_s \| C_L, C_R, \lambda, A, G) = \frac{P(X_s \| C_L, \lambda, A, G) P(X_s \| C_R, \lambda, A, G)}{P(X_s \| \lambda, A, G)}$ (Using junction tree*)	• LVCSR • Clean accented speech • ATR accented EDB (~ 80 hours of speech) • ATR accented EDB • Up to 12.7% relative WER reduction • Significant (Sign test $\alpha = 0.05$) $p \leq 0.000286$

*Resulted from equivalent BN topology (see Section 3.2.3)

A comparison of incorporating various knowledge sources at different levels of the HMM is outlined in Figure 4.38. Four different systems are included: triphone HMM baseline, pentaphone HMM baseline, and proposed models incorporating knowledge sources at HMM state and phonetic unit levels. In order to avoid decoding complexity, the pentaphone HMM baseline and the proposed pentaphone C1L3R3 were implemented by rescoring the N-best list generated from a standard and unmodified triphone ASR system. All models had five mixture components per state and incorporated knowledge sources of either phonetic-context information only or a combination of accent, gender and phonetic context information.

In the case of context dependencies only, the pentaphone HMM baseline produced only slight improvement over the triphone HMM baseline. This may be due to the sparseness of training data. The resolution of the context has been reduced, since there seemed to be too many different pentaphone contexts sharing the same Gaussian components. Thus, approximating a pentaphone model using the proposed Bayesian pentaphone C1L3R3 could help to reduce the loss of context resolution and improve performance. However, by incorporating the pentaphone knowledge in the triphone state PDF by means of a BN, a larger improvement could be obtained. The best performance that was achieved was a word accuracy of 85.03% with the proposed fLRC-HMM/BN model.

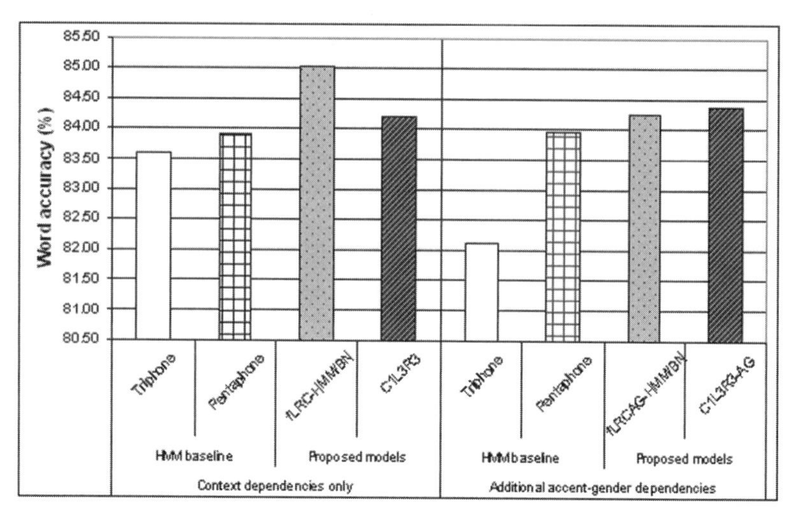

Fig. 4.38. Comparing recognition accuracy rates of different systems: triphone HMM baseline, pentaphone HMM baseline, and proposed models incorporating knowledge sources at HMM state and phonetic unit levels.

In the case of incorporating a combination of accent, gender and wide-phonetic information, the triphone HMM baseline could not achieve optimal performance. Its performance even decreased to a word accuracy of 82.11%. This might be due to the amount of accent and gender dependent training data, which was much smaller compared to the previous baseline model. As can be seen, we obtained improved recognition with fLRCAG-HMM/BN by only changing the probability distribution of states to incorporate various type of knowledge sources using BN. However, this performance of fLRCAG-HMM/BN was not better than the recognition rate of the fLRC-HMM/BN. This problem may again be related to the limited amount of training data for each accent or gender dependent model. The performance of the pentaphone HMM baseline and the proposed pentaphone C1L3R3-AG did not decrease when gender and accent were incorporated, which is probably due to the use of interpolation in the rescoring process. The best performance that was achieved was a word accuracy of 84.38% with the proposed C1L3R3-AG model.

These results reveal that additional knowledge sources of the second preceding context C_L and succeeding context C_R were appropriate to incorporate at the HMM state level, while additional knowledge sources of accent A and gender G were more appropriate to incorporate at the HMM phonetic model level. One reason may be that within one utterance the wide phonetic context information (C_L and C_R) often changes over time and thus is more appropriately incorporated at the lower level, while accent A and gender G are more constant and thus more appropriately to be incorporated at a higher level.

5

Conclusions and Future Directions

In this last chapter, we draw our conclusions and discuss future directions toward developing a spoken language dialog system.

5.1 Conclusions

This book offer a solution to enhance the robustness of a statistical automatic speech recognition system by incorporating various additional knowledge sources while keeping the training and recognition effort feasible. A new unified framework has been proposed and the efficiency of its usage has also been analyzed experimentally. A review of our work covering theoretical, application, and experimental issues is given in the following sections.

5.1.1 Theoretical Issues

Despite the rapid progress made in statistical speech recognition, many challenges still need to be overcome before ASR systems can reach their full potential in widespread everyday use.

We learned that only a limited level of success could be achieved by relying solely on statistical models while ignoring additional knowledge sources that may be available. However, there have often been cases where developing complex models and achieving optimal performance have not been simultaneously feasible. This especially applies when there are insufficient resources, i.e., the amount of training data and memory space available, for proper training of model parameters. As a result, input-space resolution may be lost due to non-robust estimates and the increasing number of unseen patterns. Moreover, decoding with large models may also become cumbersome and sometimes even impossible.

Therefore, we developed a new efficient framework to incorporate knowledge sources that can maintain a delicate balance between the ability to obtain

S. Sakti et al., *Incorporating Knowledge Sources into Statistical Speech Recognition*,
Lecture Notes in Electrical Engineering 42, DOI: 10.1007/978-0-387-85830-2_5,

deep knowledge and the feasibility of the training and recognition effort. This proposed framework, GFIKS, was designed by utilizing the BN framework, and it was formulated to answer some key problems:

- How can we incorporate various knowledge sources from different domains in an efficient and unified way?
- At which level of the ASR system can we incorporate these additional knowledge sources?
- How can we solve model complexity when incorporating a significant number of additional knowledge sources? Issues discussed include the availability of training data as well as the training and recognition efforts.

Our framework allows probabilistic relationships among different information sources to be learned. Therefore, various kinds of knowledge sources can be incorporated, and a probabilistic function of the model can be formulated. This framework was structured in a general way so that it could be applied to many existing ASR problems with their respective model-based likelihood functions, at any level and in flexible ways. Based on the BN framework, the proposed framework also facilitates the decomposition of the joint PDF into a linked set of local conditional PDFs using a junction tree algorithm. Consequently, a simplified form of the model can be constructed and reliably estimated using a limited amount of training data. In Chapter 3, we briefly described the basic theory of BN and formulated the design of our proposed general framework.

5.1.2 Application Issues

We examined the use of the proposed framework by incorporating various additional knowledge sources at different levels of an acoustic model.

We first demonstrated in Section 4.1 how the additional sources of knowledge are incorporated at the HMM state level. This was done by incorporating noise, accent, gender and wide-phonetic context information, which often suffers from a sparseness of data and memory constraints, in a triphone HMM by means of BN. At this level, since the state output probability can still be easily modeled with a Gaussian function, we can thus directly infer state output from the BN. If the additional knowledge sources are assumed to be hidden during recognition, our approach allows the use of the standard decoding system without modification.

We then demonstrated how additional knowledge sources are incorporated at a higher level, the HMM-phonetic-model level, in Section 4.2. In contrast to the state level, obtaining a simple functional form is difficult for the conditional PDF here because it involves an HMM model and a speech segment of variable duration. Therefore, we applied the form of the junction tree algorithm where BN directed graphs are decomposed into clusters of variables on which relevant computations can be carried out. As a result, a wide-phonetic context HMM model is constructed through composition of several other models with narrower contexts. Because this composition technique reduces the number of context units to be estimated, the resolution of contexts was considerably improved since only the less context-dependent models needed to be estimated. To simplify the decoding mechanism, we applied these wide-context-model compositions at the post-processing stage with N-best rescoring.

5.1.3 Experimental Issues

We verified the proposed approaches in various English LVCSR tasks, using accented or noisy speech data. These were tested under matching and mismatched conditions.

At the HMM-state level, we first incorporated gender, noise and wide-phonetic context information independently. The proposed HMM/BN models always performed better than did the triphone HMM baseline within the same number of parameters. By changing only the probability distribution, we improved recognition. For noisy speech on the AURORA task, The proposed HMM/BN performs even better than DBN, especially at low SNRs. We then incorporated together three different types of knowledge sources: accent, gender and wide-phonetic context. However, compared with the incorporation of wide-phonetic context only, the additional knowledge sources of gender and accent variables did not improve the recognition rate of the proposed models any further. This problem may be related to the limited amount of training data for each accent- or gender-dependent pentaphone model. Consequently, the best performance was obtained by incorporating only the wide-phonetic context information.

Furthermore, at the HMM-phonetic-model level, by incorporating wide-phonetic context information, the proposed Bayesian pentaphone model, C1L3R3, always improved performance relative to the baseline, and this was better than simply rescoring with a conventional pentaphone HMM. The context resolution of a conventional pentaphone HMM was reduced, since there seemed to be too many different pentaphone contexts sharing identical Gaussian components. Thus, approximating a pentaphone model using the composition of several less-context-dependent models, such as the proposed Bayesian pentaphone C1L3R3, could help increase the resolution of context and improve performance. The performance also further improved when we combined the model with other knowledge sources, including gender and accent information. This was in contrast to the case using pentaphone HMM/BN, which probably reflects the use of deleted interpolation. The best performance was achieved with a C1L3R3-AG that incorporated additional knowledge of accent A, gender G, second preceding context C_L, and second succeeding context C_R.

In summary, these experimental results revealed that wide-phonetic context models, which were developed with the proposed framework, improved word accuracy in comparison to standard HMM models with and without additional knowledge sources. Additional knowledge of the second preceding context, C_L, and that of the second succeeding context, C_R, were appropriate to incorporate at the HMM state level; on the other hand, additional knowledge of accent A and that of gender G were more appropriate to incorporate at the HMM-phonetic-model level.

5.2 Future Directions: A Roadmap to a Spoken Language Dialog System

The future development of the presented ASR system includes its implementation in a spoken language dialog system. Even though the recognition system described in this book offers many advances and successes, there is still room for further enhancement.

If speech recognition systems were capable of interacting with humans, it would seem reasonable for them to make use of the many modalities inherent in human communication, not only the acoustic-phonetic variabilities of speech but also other speaker factors and emotions. According to Quast (2001), information in the speech signal include:

1. Verbal content

   Verbal content deals with the actual meaning of the spoken words or the linguistic part, i.e., coarticulation (neighboring phonetic context).

2. Non-verbal content

   Non-verbal content deals with the following two parts:
   a) Extralinguistic

      The speaker's basic state, i.e., gender, accents, and so on.

   b) Paralinguistic

      The speaker's expression of emotions, i.e., the prosody.

In this book, we presented applications for incorporating wide-phonetic context information as well as gender and accent information, considering the wide-phonetic context information as linguistic content and the gender-accent information as extralinguistic content. Accordingly, further investigations that incorporate other knowledge sources would be beneficial, especially the knowledge sources of paralinguistic content, i.e., the prosody.

Toward the goal of developing a spontaneous ASR technology, ASR must have the capability to capture wide knowledge sources on different ASR levels, including:

1. State level (HMM state)
2. Phonetic unit level (HMM phonetic unit model)
3. Word level (Pronunciation lexicon)
4. Sentence level (Language model)

In this book, we focused on acoustic models and demonstrated knowledge incorporation at the HMM-states and phonetic-model levels. Consequently, further investigations at higher ASR levels, i.e., a pronunciation lexicon and language model, are also important.

The roadmap to a spoken-language dialog system is outlined in Figure 5.1, highlighting further investigations that incorporate other knowledge sources at other ASR levels. Our goal is advance toward higher ASR levels by using more challenging knowledge sources.

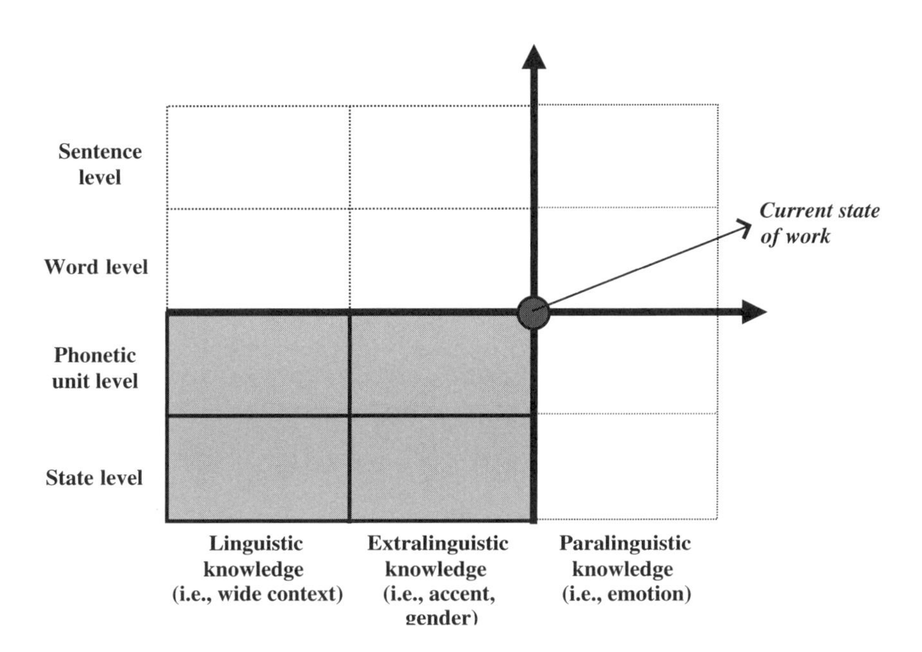

Fig. 5.1. Roadmap to spoken language dialog system incorporating other knowledge sources at higher ASR levels.

A

Speech Materials

A.1 AURORA TIDigit Corpus

The AURORA 2000 noisy-speech speaker-independent digits task was generated by Ericsson Eurolab, as a contribution to the ETSI STQ-AURORA DSR Working Group (Hirsch and Pearce, 2000). AURORA is developing standards for distributed speech recognition (DSR), where the speech analysis is done in the telecommunication terminal and the recognition is done at a central location in the telecom network. It belongs to the technical body STQ (Speech processing, Transmission and Quality aspects) as ETSI standardization activity. The database has been publicly available and widely used by speech researchers as a standard database to evaluate the performance of speech-recognition systems. The source speech of this database is TIDigits, consisting of a connected-digits task using the following eleven words: "zero," "oh," "one," "two," "three," "four," "five," "six," "seven," "eight," "nine," which are spoken by American English talkers.

TIDigits is taken as basis, since its data are considered as "clean." Here, additional filtering techniques, G.172 and MIRS, are applied to consider the realistic frequency characteristics of terminals and equipment in the telecommunications domain. Then, the distortions are artificially added at a desired SNR, where the noise signals are selected to represent the most probable application scenarios for telecommunication terminals, such as:

- Suburban train
- Babble (crowd of people)
- Car
- Exhibition hall
- Restaurant
- Street
- Airport
- Train station

The database is separated into two sets :

- Training set

  This consists of 8,440 utterances spoken by 55 female and 55 male speakers and filtered with G.172 frequency characteristics. These utterances, which consist of about one half million frames or 4.167 hours of speech time, are equally split into 20 subsets with 422 utterances in each subset. These 20 subsets represent four different noise scenarios: subway (suburban train), babble (crowd of people), car and exhibition hall, at 5 different SNRs (20 dB, 15 dB, 10 dB, 5 dB, and clean condition).

- Testing set

  The test sets consist of 4,004 utterances from another 52 males and 52 female speakers, which are equally split into four subsets with 1,001 utterances in each. One noise signal is added to each subset of 1,001 utterances at four different SNRs (20 dB, 10 dB, 0 dB, and clean condition). One test set includes a total of 16,016 files, more or less twice the size of the training set (about 8 hours of speech time). These test sets are:

  1. Test set A

     This test set leads to a high matching of training data, with the same G.172 filter characteristics and the same kind of noise scenario (suburban train, babble, car and exhibition hall).

  2. Test set B

     This test set shows the influence on recognition when considering different noises than those used for training. The additive noise scenario is mismatched (restaurant, street, airport, and train station).

  3. Test set C

     This test set shows the influence on recognition performance when a different frequency is present at the input of the recognizer. Here, only two of the four subsets are used, and they are filtered with MIRS characteristics.

Note that, in this research work, we evaluated only test sets A and B.

A.2 TIMIT Acoustic-Phonetic Speech Corpus

The DARPA TIMIT Acoustic-Phonetic Continuous Speech Corpus (Zue et al., 1990) was designed in a joint effort among the Massachusetts Institute of Technology (MIT), SRI International (SRI), and Texas Instruments (TI) to provide speech data for the acquisition of acoustic-phonetic knowledge and for the development and evaluation of automatic speech-recognition systems.

The TIMIT corpus contains a total of 6300 sentences, 10 sentences spoken by each of 630 speakers from 8 major dialect regions of the United States. Table A.1 shows the number of speakers for the 8 dialect regions (dr), includes:

- dr1: New England
- dr2: Northern
- dr3: North Midland
- dr4: South Midland
- dr5: Southern
- dr6: New York City
- dr7: Western
- dr8: Army Brat (moved around)

Table A.1. *Dialect distribution of speakers.*

Dialect region (dr)	# Male speaker	# Female speaker	Total
1	31 (63%)	18 (27%)	49 (8%)
2	71 (70%)	31 (30%)	102 (16%)
3	79 (67%)	23 (23%)	102 (16%)
4	69 (69%)	31 (31%)	100 (16%)
5	62 (63%)	36 (37%)	98 (16%)
6	30 (65%)	16 (35%)	46 (7%)
7	74 (74%)	26 (26%)	100 (16%)
8	22 (67%)	11 (33%)	33 (5%)
Total	438 (70%)	192 (30%)	630 (100%)

The completed text material in the TIMIT prompts consists of 2 dialect "shibboleth" sentences designed at SRI, 450 phonetically compact sentences designed at MIT, and 1890 phonetically diverse sentences selected at TI. The phonetically compact sentences were designed to provide a good coverage of pairs of phones, with extra occurrences of phonetic contexts thought to be either difficult or of particular interest. The phonetically diverse sentences (the SI sentences) were selected from existing text sources - the Brown Corpus (Kuchera and Francis, 1967) and the Playwrights Dialog (Hultzen et al., 1964) - so as to add diversity in sentence types and phonetic contexts.

The dialect sentences (the SA sentences) are meant to reveal the dialectal variants of the speakers and are read by all 630 speakers. For the phonetically-compact sentences (the SX sentences), each sentence are read by seven different speakers, while each speaker read five of these sentences. The phonetically diverse sentences (the SI sentences) are read by only a single speaker and each speaker read three of these sentences. The total number of utterances is summarized in Table A.2.

Table A.2. *Speech materials of TIMIT database.*

Type	Sentence	Speakers	Total	Sentences/speaker
Dialect (SA)	2	630	1260	2
Compact (SX)	450	7	3150	5
Diverse (SI)	1890	1	1890	3
Total	2342		6300	10

Each sentence in the TIMIT database has a time-aligned phoneme transcription. Table A.3 contains detailed statistics of the subset of the TIMIT comprised of the SX and SI sentences.

Table A.3. *Statistics on the TIMIT database.*

	SX and SI set
Number of sentences	5040
Number of words	41161
Number of unique words	5107
Average number of words per sentence	8.2
Average number of syllables per word	1.5
Average number of phones per words	3.9

A.3 Wall Street Journal Corpus

The well-known continuous-speech recognition (CSR) Wall Street Journal (WSJ) corpus (Paul and Baker, 1992) was designed by the DARPA CSR Corpus Coordinating Committee (CCCC) and collected at the MIT, SRI International, and TI.

The corpus materials consist primarily of native English (North American) read speech with texts drawn from a machine-readable corpus of Wall Street Journal news text. The articles were quality filtered to limit the vocabulary to the 64,000 most frequently occurring words in the database, which consists of approximately 37 million words of text.

The corpus is scalable and built to accommodate variable size large vocabulary, speaker-dependent (SD) and -independent (SI) training with variable amounts of data (ranging from 100 to 9,600 sentences/speaker), including equal portions of verbalized and non-verbalized punctuation (to reflect both dictation mode and non-dictation mode application), a simultaneous standard close-talking Sennheiser HMD414 microphone and multiple secondary microphones, and equal numbers of male and female speakers chosen for diversity of voice quality and dialect.

The collection of the database was performed in two phases (Odell, 1995) as described in the following:

- CSR-WSJ0

  The complete WSJ0 corpus was split almost equally into three sections:
  - Longitudinal speaker dependent, *LSD*
  - Long-term speaker independent, *SI12*
  - Short-term speaker independent, *SI84*

  In this study, only the short-term speaker-independent *SI84* section was used for training. It consists of 7,240 sentences from 84 different speakers (42 males and 42 females) for a total of around 15.3 hours of speech.

- CSR-WSJ1

  The CSR-WSJ1 corpus was split almost equally into two sections:
  - Long-term speaker independent, *SI25*
  - Short-term speaker independent, *SI200*

  Again, of these only the short-term speaker independent *SI200* section was used. It consists of 29,320 sentences from 200 new speakers (100 males and 100 females) for a total 45.1 hours of speech.

The combination of both short-term speaker-independent portions of the database, *SI84* and *SI200*, are collectively referred to as *SI284*, with more than 60 hours of speech in total.

The WSJ corpus includes several tests, known as the Hub and Spoke evaluation paradigm. The entire test suite for evaluation consists of two Hub tests and nine Spoke tests. In this research work, only the smaller 5k Hub test was used. It was created by using prompting texts from the 5k-word test pools specified in the WSJ0 corpus. These articles were filtered to discard paragraphs with more than one word outside of the 5k most frequent words in the corpus.

A.4 ATR Basic Travel Expression Corpus

The basic travel expression corpus (BTEC) described in (Kikui et al., 2003) was planned to cover utterances for every potential subject in travel conversations, together with their translations. Since it is rather infeasible to collect them through transcribing actual conversations or simulated dialogues, ATR decided to use sentences from the advice of bilingual travel experts, by investigating the "phrasebooks" containing Japanese/English sentence pairs that those experts consider useful for tourists traveling abroad.

Currently, BTEC serves as the primary source for developing broad-coverage speech-to-speech translation (S2ST). It is now being translated into several languages including Chinese, French, German, Italian and Korean by members of the C-STAR (International Consortium for Speech Translation Advanced Research). In addition, it will be further translated into Indonesian, Thai, and Hindi by members of the A-STAR (International Consortium of Speech Translation Advanced Research in Asia).

The full BTEC test set1 consists of 4,080 read speech utterances spoken by 40 different speakers (20 males, 20 females). In this study, in order to reduce the recognition time, we simply selected 1,000 utterances spoken by 20 different speakers (10 males, 10 females), and used them as a development set. Two hundred randomly selected utterances spoken by 40 different speakers (20 males, 20 females) were used as a test set.

A.5 ATR English Database Corpus

The ATR English database (EDB) corpus was designed for speech-to-speech translation in the travel conversation task. The text sentences were collected from various different corpora as follows:

1. Basic travel expression corpus

   See the description of BTEC in A.4.

2. TIMIT acoustic-phonetic continuous speech corpus

   See the description of TIMIT in A.2.

3. Machine translation aided dialogue corpus

   The machine translation aided dialogue (MAD) corpus described in (Kikui et al., 2003) was designed by carrying out simulated (i.e. role play) dialogues between two native speakers of different mother tongues with a

Japanese/English bi-directional S2ST system, instead of using human interpreters. In order to concentrate on the effects of machine translation (MT) by circumventing communication problems caused by speech recognition errors, the speech recognition modules were replaced with human typists. The resulting system is thus considered equivalent to using an S2ST system whose speech recognition part is nearly perfect.

The number of sentences collected from each corpus is summarized in Table A.4. In addition, 150 single-word sentences of place names are also included. Based on these text materials, the speech corpora were collected from accented English speakers: "American" (US), "British" (BRT), and "Australian" (AUS). The total number of speakers and utterances, as well as speech hours, are summarized in Table A.5.

Table A.4. *Text sentence materials of ATR English speech database.*

Corpus	# Unique sentences
BTEC	2400
MAD	535
TIMIT	2342
Single words	150
Total	5427

Table A.5. *Speech materials of ATR English speech database.*

Accent	# Speakers (M,F)	# Utterances	# Hour of speech
US	199 (99, 100)	91,484	88.9
BRT	100 (50, 50)	45,532	44.3
AUS	100 (50, 50)	45,934	44.7
Total	399 (199, 200)	182,950	177.9

In this study, in order to reduce the training time, we simply selected US and AUS accented speech only, with about 45,000 utterances ($\sim$ 44 speech hours) spoken by 100 speakers (50 males and 50 females) for each accent. We used 90% of the data (20,000 utterances spoken by 40 speakers for each US male, US female, AUS male and AUS female) or about 80,000 utterances in total ($\sim$ 80 speech hours) as the training data.

For testing data of the accented read speech evaluation, we randomly selected 200 utterances, spoken by 20 different speakers (10 males and 10 females) from the remaining 10% of the mixed-accent data (US and AUS) of the ATR-EDB corpus.

B

ATR Software Tools

This chapter describes a brief outline of ATR software tools, the so-called "ATRASR", and the way that they are used to construct and test the speech recognizers.

B.1 Generic Properties of ATRASR

ATRASR consists of several functionality modules which are designed to run with a traditional command-line style interface. Each module has a number of required arguments plus optional arguments. As well as providing a consistent interface, ATRASR support also multiple file formats allowing input and output data with different file formats. Several modules can also be compiled into one new module.

In addition to command line arguments, the operation of a tool can be controlled by parameters stored in a configuration file. In this case, the ATRASR will load the parameters stored in the configuration file config during its initialization procedures.

A description of any ATRASR modules can be obtained simply by executing the module with no arguments.

B.2 Data Preparation

In order to create an acoustic model, the following data files are required:

1. Pronunciation dictionary (lexicon)

   This is a sorted list of the words including their pronunciations. The ATRASR lexicon dictionary format for phone-based model is :

   ```
   <ID>  [<Output Symbol>]  <phone1>  <phone2> ... {|SIL}
   ```

It uses $|S|$ as "START" symbol and $|E|$ as "END" symbol which has a silence model SIL as their pronunciation. Here is an example of AURORA2 phone-based lexicon dictionary:

```
|E| []       SIL
|S| []       SIL
ONE [ONE]   W AH N {|SIL}
TWO [TWO]   T UW {|SIL}
THREE [THREE]   TH R IY {|SIL}
FOUR [FOUR]   F AO R {|SIL}
FIVE [FIVE]   F AY V {|SIL}
SIX [SIX]   S IH K S {|SIL}
SEVEN [SEVEN]   S EH V AX N {|SIL}
EIGHT [EIGHT]   EY T {|SIL}
NINE [NINE]   N AY N {|SIL}
ZERO [ZERO]   Z IH R OW {|SIL}
ZERO [ZERO]   Z IY R OW {|SIL}
OH [OH]   OW {|SIL}
```

2. A set of speech data files

   This is composed a speech database of recorded human utterances. ATR mainly uses 16-kHz sampling frequency, 16-bit linear quantization, big endian format and raw files.

3. A set of transcription files

   To create a set of HMM-net acoustic modeling, both speech and text database are necessary. Where the text database (transcription) is used as an answer during learning process of speech data. In this case, every file of speech data ("WAV" files training set) must have an association with transcription text. There are several transcription format used during training. One type of transcription file consists of acoustic-unit level transcription with speech utterance starting time and ending time for the whole utterance. These files are usually used during the embedded training process. The ATRASR "TRS" file format is:

```
<start time [ms]> <phone sequence> <end time [ms]>
```

   For example, here is "FAC_6O46A.TRS" in AURORA2 database :

```
0.000000  SIL,S,IH,K,S,OW,F,AO,R,S,IH,K,S,SIL     2.086375
```

   The utterance starts at 0.000000 ms and ends at 2.086375 ms.

Another type of transcription file consists of acoustic-unit level transcription, where it gives time information for each phoneme called ATRASR "SLF" file. These files will be used during the acoustic-unit topology and label training process.

4. A set of text files

   These text files are needed for language modeling training (see Section B.5).

B.3 SSS Data Generating Tools

Before the training process, the utterances of speech data files must be segmented into phoneme and word labels according to lexicon dictionary and the speech association transcription text. This is done by forced alignment using *ATR_trace* module.

Then, the segmented utterances are converted into appropriate format, where each segment should supplies information about phoneme context sequences followed by floating-point binary of the MFCC feature parameter. The general structure of phone-based SSS data is as follows :

```
<speaker ID>
<# of phonemes>
<phone sequences>
<# of frames>
[Floating point binary data of MFCC feature parameter]
```

This performed using *ATR_labelSSS* and *ATR_embSSS* modules for each phoneme sequence and each utterance phoneme sequences, respectively. An example process of generating phoneme-based SSS data is shown in Figure B.1.

B.4 Acoustic Model Training Tools

The creation of HMM-net acoustic modeling basically consists of four following steps:

1. Step 1:

   The first step of training process is to define the topology required for each HMM, where in our case is each phoneme acoustic model (without silence model).

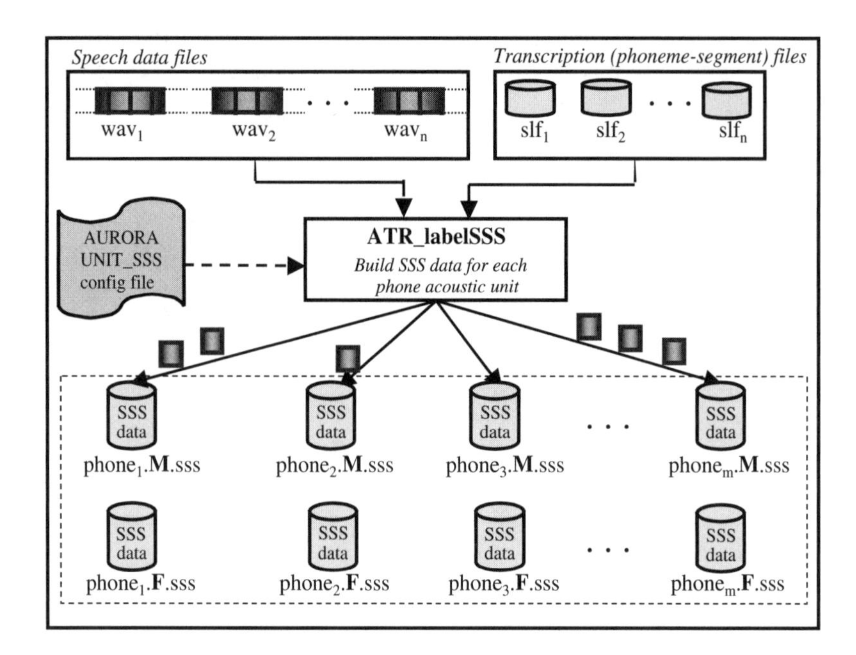

Fig. B.1. The ATRASR phoneme-based SSS data creation for phone-unit model training.

ATRASR construct HMMs topology automatically based on MDL-SSS algorithm (Jitsuhiro et al., 2004) in order to give an optimum number of states for acoustic modeling. This process is performed with ATRASR tool called *ATR_trainHMnet*.

Then, each model parameters are trained using the data described in previous section. This process is performed with *ATR_retrainHMnet*.

2. Step 2:

   The silence acoustic model is constructed with a fixed standard HMM topology. Two types of silence acoustic models: 1-state and 3-state models are generated.

3. Step 3:

   In third step, all phoneme acoustic models are combined into one embedded acoustic model. Then the 1-state silence acoustic model is added into whole embedded acoustic model.

This model is then retrained again in order to give a good recognition performance on one complete utterance. This process was performed with *ATR_retrainHMnet*.

4. Step 4:

   At last, the 1-state silence acoustic model is replaced with 3-state silence acoustic model.

The topology training for each phone acoustic unit is illustrated in Figure B.2, and the whole embedded training is illustrated in Figure B.3.

B.5 Language Model Training Tools

A language model for speech recognition is usually trained using a large amounts of text data, and the process consists of following steps:

1. Step 1:

   The first step in building a language model is counting the n-grams in your training text using *ATR_L2train*

2. Step 2:

   The next step is constructs the actual language model using *ATR_L2smooth* from the resulted count files.

Note that, in some cases, in order to cope with the data sparseness problem, class-based N-gram model was proposed. In this case, instead of dealing with separated words, class-based N-gram estimates parameters for word classes.

B.6 Recognition Tools

To perform the Viterbi-based speech recognition, the following inputs are required:

1. A set of speech files
2. A set of transcription files
3. An acoustic model
4. A pronunciation lexicon
5. A language model

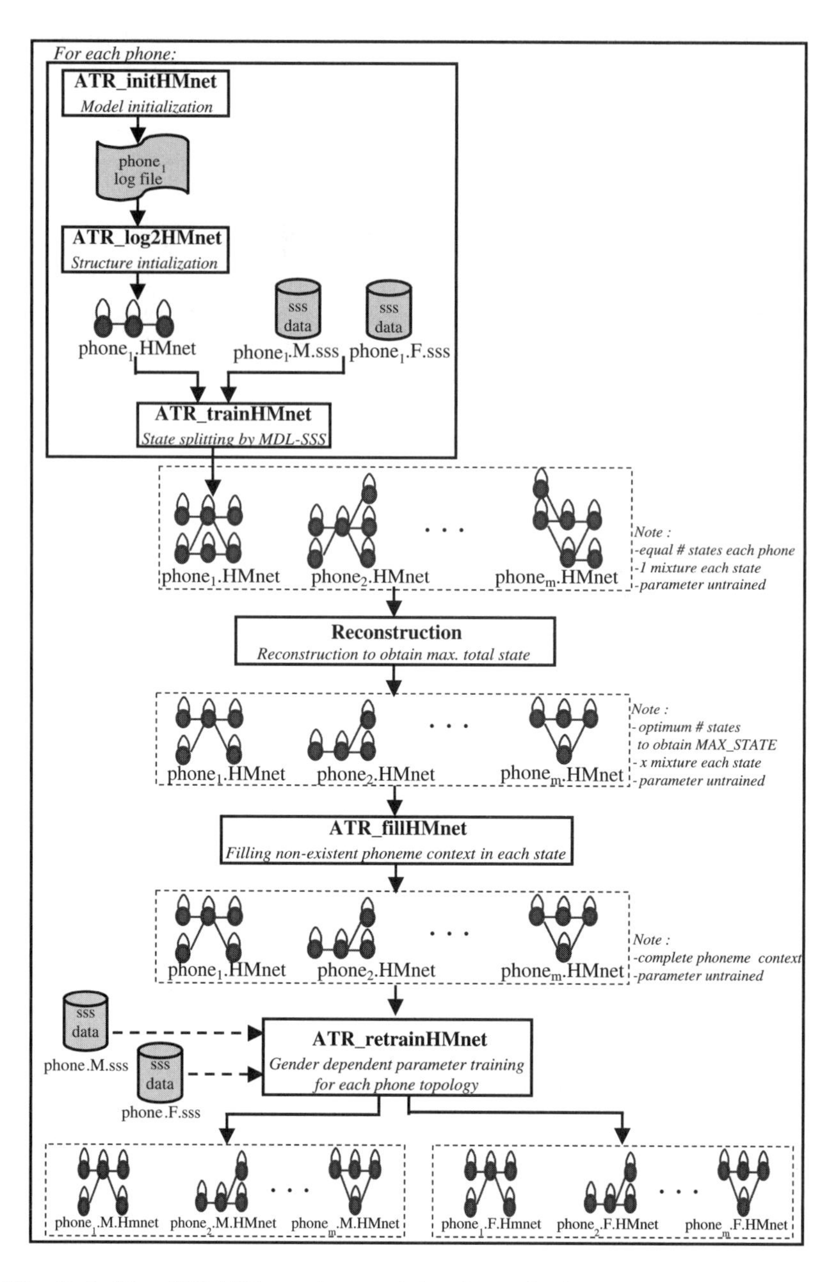

Fig. B.2. The ATRASR topology training for each phone acoustic-unit model.

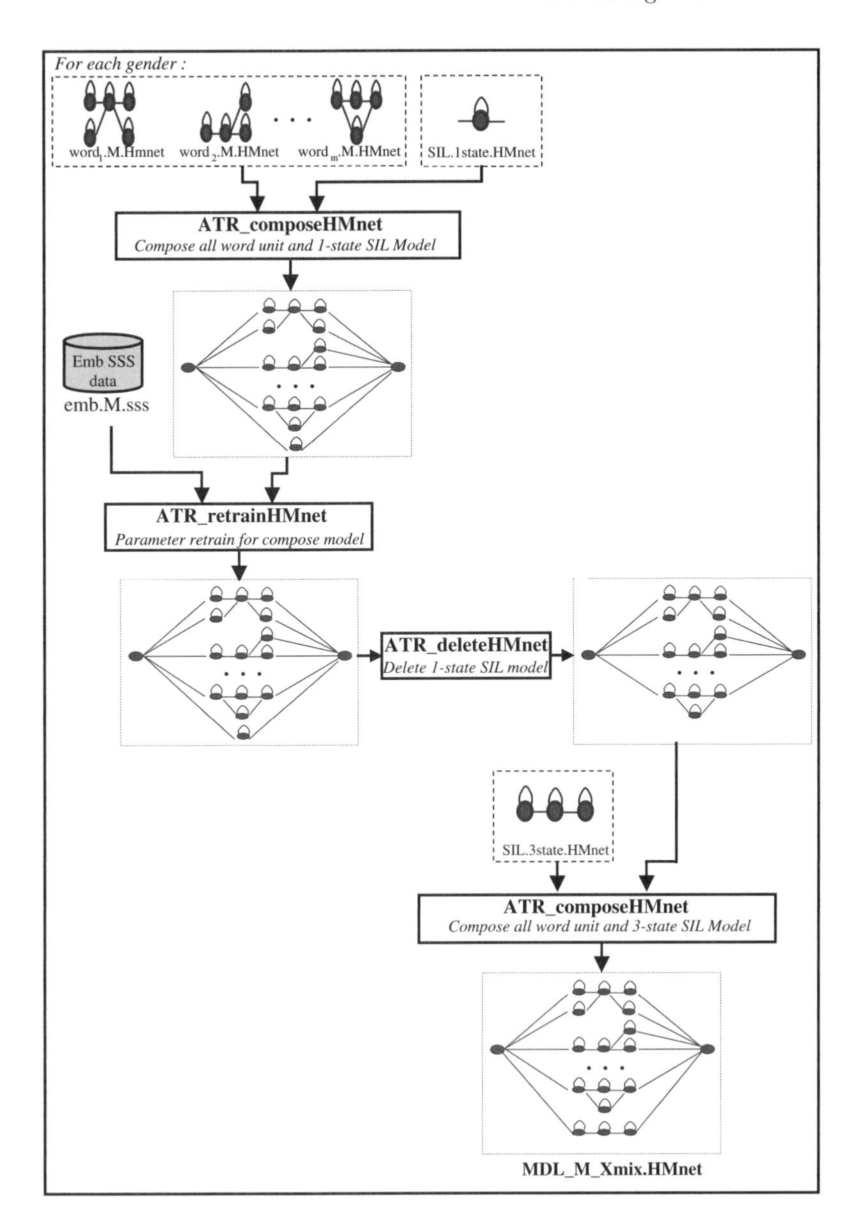

Fig. B.3. The ATRASR embedded training for a whole HMnet.

This recognition process, which is illustrated in Figure B.4, is performed using *ATR_wcclrr* module, which is composed from several modules, including:

- *ATR_wave2para* to estimate MFCC feature parameter from WAV file of test data,
- *ATR_lattice* to generate lattice word network and calculate the probability using bigram language model,
- *ATR_rescore* to rescore the probability with trigram language model, and *ATR_result* to calculate the word error rate.

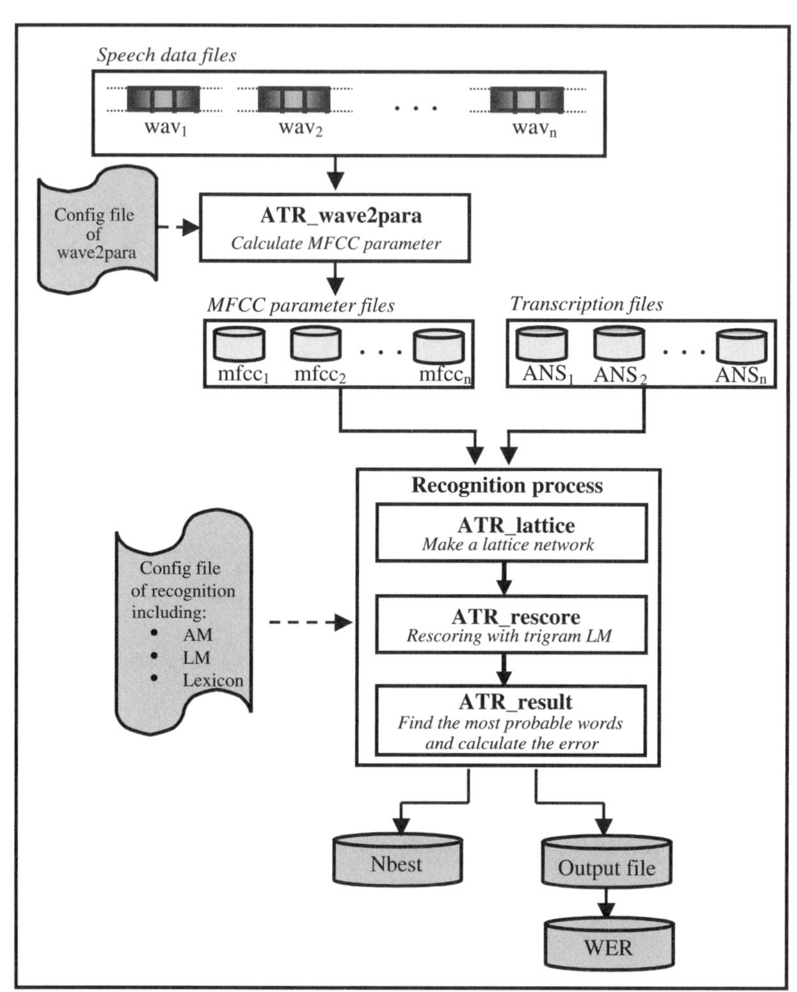

Fig. B.4. The recognition process using ATRASR tools.

An example of a word error rate summary in the output file is described as follows:

```
# total utterances = 1001

# best accuracy    = 98.001998, perfect = 981

# net accuracy     = 99.400599, perfect = 995

# total words = 3257

# best accuracy = 99.355235, INS = 8, DEL = 6, SUB = 7

# best correct = 99.600860, DEL = 6, SUB = 7

# net accuracy = 99.815781, INS = 0, DEL = 5, SUB = 1

# net correct = 99.846484, DEL = 4, SUB = 1
```

It means that from 1001 sentences which consists of 3257 words, there are 8 insertion (INS), 6 deletion (DEL), and 7 substitution (SUB). So it gives the best accuracy 99.36%.

A more detailed description of the ATR software tools can be found in (Nakamura et al., 2006; Sakti, 2005).

C

Composition of Bayesian Wide-phonetic Context

As described in Section 4.3.2, a conditional probability function of a wide-phonetic context model may be decomposed of several less complex models, using junction tree algorithm, as

$$P(X_s|C_L, \lambda, C_R) = \frac{P(X_s|C_L, \lambda)P(X_s|\lambda, C_R)}{P(X_s|\lambda)}. \tag{C.1}$$

In the case where we assume that λ is monophone unit $/a/$, and C_L and C_R are the ones preceding and following context unit $/a^-/$ and $/a^+/$, respectively, we can obtain the same factorization as one that has been proposed by Ming et al. (1999); Ming and Smith (1998), and that is known as the Bayesian triphone. The extended version of the Bayesian triphone, the so-called Bayesian wide-phonetic context model, can also be found in our previous study (Sakti et al., 2006, 2005). This approach allows us to model a wide range of phonetic contexts from less context-dependent models simply based on Bayes' rule.

In this chapter, we perform the proof using the Bayesian principle (Section C.1), and define some variants of Bayesian Wide-phonetic Context which can derived simply using Bayes' rule (Section C.2).

C.1 Proof using Bayes's Rule

Using the Bayesian principle:

$$\begin{aligned} P(X_s|C_L, \lambda, C_R) &= \frac{P(X_s, C_L, \lambda, C_R)}{P(C_L, \lambda, C_R)} \\ &= \frac{P(C_L, C_R|\lambda, X_s)P(\lambda, X_s)}{P(C_L, C_R|\lambda)P(\lambda)}. \end{aligned} \tag{C.2}$$

Assuming that C_L and C_R are independent given λ and X_s, $P(C_L, C_R|\lambda) = P(C_L|\lambda)P(C_R|\lambda)$, $P(C_L, C_R|\lambda, X_s) = P(C_L|\lambda, X_s)P(C_R|\lambda, X_s)$, and Eq. (C.2) becomes:

$$P(X_s|C_L, \lambda, C_R) = \frac{P(C_L|\lambda, X_s)P(C_R|\lambda, X_s)P(\lambda, X_s)}{P(C_L|\lambda)P(C_R|\lambda)P(\lambda)}. \tag{C.3}$$

By multiplying both the numerator and denominator by $P(\lambda, X_s)P(\lambda)$ and applying Bayes's rule, Eq. (C.3) becomes:

$$\begin{aligned} & P(X_s|C_L, \lambda, C_R) \\ &= \frac{P(C_L|\lambda, X_s)P(\lambda, X_s)}{P(C_L|\lambda)P(\lambda)} \frac{P(C_R|\lambda, X_s)P(\lambda, X_s)}{P(C_R|\lambda)P(\lambda)} \frac{P(\lambda)}{P(\lambda, X_s)} \\ &= \frac{P(X_s|C_L, \lambda)P(C_L, \lambda)}{P(C_L, \lambda)} \frac{P(X_s|\lambda, C_R)P(\lambda, C_R)}{P(\lambda, C_R)} \frac{1}{P(X_s|\lambda)} \\ &= \frac{P(X_s|C_L, \lambda)P(X_s|\lambda, C_R)}{P(X_s|\lambda)}, \end{aligned} \tag{C.4}$$

which is the same result as in Eq. (C.1).

As can be seen, a wide range of phonetic contexts may be decomposed from less context-dependent models simply based on Bayes' rule. However, difficulties arise when different types of knowledge sources need to be incorporated. In contrast, the current unified framework gives us a more appropriate means of incorporating various kinds of knowledge sources. For example, we can easily extend further C1L3R3 with other additional knowledge variables, such as gender or accent information (See Section 4.3.2).

C.2 Variants of Bayesian Wide-phonetic Context Model

Using Eq. (C.4), we set λ to be a triphone unit $/a^-, a, a^+/$, C_L to be the second preceding context $/a^{--}/$, and C_R to be the second following context $/a^{++}/$, then $P(X_s|C_L, \lambda, C_R)$ will become the C3L4R4 composition as derived in Eq. (4.26)

$$P(X_s|C_L, \lambda, C_R) = \frac{P(X_s|[a^{--}, a^-, a, a^+])P(X_s|[a^-, a, a^+, a^{++}])}{P(X_s|[a^-, a, a^+])}. \tag{C.5}$$

By approximating the probability distribution of composition C3L4R4 with more reduced models, such as $P(X_s|[a^{--}, a^-, a, a^+])$ with $P(X_s|[a^{--}, a, a^+])$, $P(X_s|[a^-, a, a^+, a^{++}])$ with $P(X_s|[a^-, a, a^{++}])$, and $P(X_s|[a^-, a, a^+])$ with $P(X_s|[a])$, Eq. (C.5) becomes:

$$\begin{aligned} P(X_s|C_L, \lambda, C_R) &= \frac{P(X_s|C_L, \lambda)P(X_s|\lambda, C_R)}{P(X_s|\lambda)} \\ &= \frac{P(X_s|[a^{--}, a^-, a, a^+])P(X_s|[a^-, a, a^+, a^{++}])}{P(X_s|[a^-, a, a^+])} \\ &= \frac{P(X_s|[a^{--}, a, a^+])P(X_s|[a^-, a, a^{++}])}{P(X_s|[a])}. \end{aligned} \tag{C.6}$$

This approximation gives another way of composing pentaphone models, where a pentaphone model is composed of $P(X|[a^{--}, a, a^+]), P(X|[a^-, a, a^{++}])$, and $P(X|[a])$. These correspond to the PDFs of the observation X given the left/preceding-skip-triphone-context (Lsk3), the right/following-skip-triphone-context (Rsk3), and the center monophone unit (C1), respectively. The composition in Eq. (C.6) is called composition C1Lsk3Rsk3.

The above algorithms, such as compositions C1L3R3, C3L4R4, and also C3Lsk3Rsk3, always follow the general representation $P(X_s|C_L, \lambda, C_R)$, where the wide-context model $/C_L, \lambda, C_R/$ is composed of left-context-dependent unit $/C_L/$, right-context-dependent unit $/C_R/$, and center base context unit $/\lambda/$. However, alternatives exist for composing wide-context models other than those described above. For example, a wide-context model is composed of several less-context-dependent models, where the center base unit $/\lambda/$ in each model is the center point of each phonetic context. Then, the PDF of X generated from the pentaphone $/a^{--}, a^-, a, a^+, a^{++}/$ context unit can be approximated as follows:

$$\begin{aligned} P(X_s|[a^{--}, a^-, a, a^+, a^{++}]) &= \frac{P(X_s, [a^{--}, a^-, a, a^+, a^{++}])}{P([a^{--}, a^-, a, a^+, a^{++}])} \\ &= \frac{P([a^{--}, a^-, a^+, a^{++}]|a, X_s)P([a], X_s)}{P([a^{--}, a^-, a^+, a^{++}]|[a])P([a])} \end{aligned}$$

$$= \frac{P([a^{--}, a^{++}]|[a], X_s)P([a^{-}, a^{+}]|[a], X_s)P([a], X_s)}{P([a^{--}, a^{++}]|[a])P([a^{-}, a^{+}]|[a])P([a])}$$

$$= \frac{P(X_s|[a^{--}, a, a^{++}])P(X_s|[a^{-}, a, a^{+}])}{P(X_s|[a])}. \quad \text{(C.7)}$$

The result indicates that a pentaphone $/a^{--}, a^{-}, a, a^{+}, a^{++}/$ model can be composed of $P(X_s|[a^{--}, a, a^{++}])$, $P(X_s|[a^{-}, a, a^{+}])$, and $P(X_s|[a])$, which correspond to the PDFs of the observation X given the center-skip-triphone-context (Csk3), center-triphone-context (C3), and center monophone unit (C1), respectively; this is called composition C1C3Csk3.

In these compositions, the number of context units to be estimated is reduced from N^5 to $(2N^3 + N)$ for compositions C1L3R3, C1Lsk3Rsk3, and C1C3Csk3, and to $(2N^4 + N^3)$ for composition C3L4R4, without loss of context coverage, where N is the number of phones. If we use a 44-phoneme set for English ASR, the total number of different contexts that need to be estimated in the pentaphone model is $44^5 =\sim 165$ million context units. Composition with triphone-context-units reduces the complexity to about 170,000 context units, but composition with tetraphone-context-units reduces the complexity to only about 7.5 million context units.

As summary, we depict Bayesian pentaphone models structure and PDF in Figure C.1, including: Bayesian C1L3R3 which is composed of the preceding/following triphone-context unit and center-monophone unit, Bayesian C3L4R4 which is composed of the preceding and following tetraphone-context unit and the center-triphone-context unit, Bayesian C1Lsk3Rsk3 which is composed of the preceding and following skip-triphone-context unit and the center-monophone unit, and Bayesian C1C3Csk3 which is composed of the center skip-triphone-context unit, the center triphone-context unit and the center-monophone unit. Here we also include the conventional pentaphone C5 model.

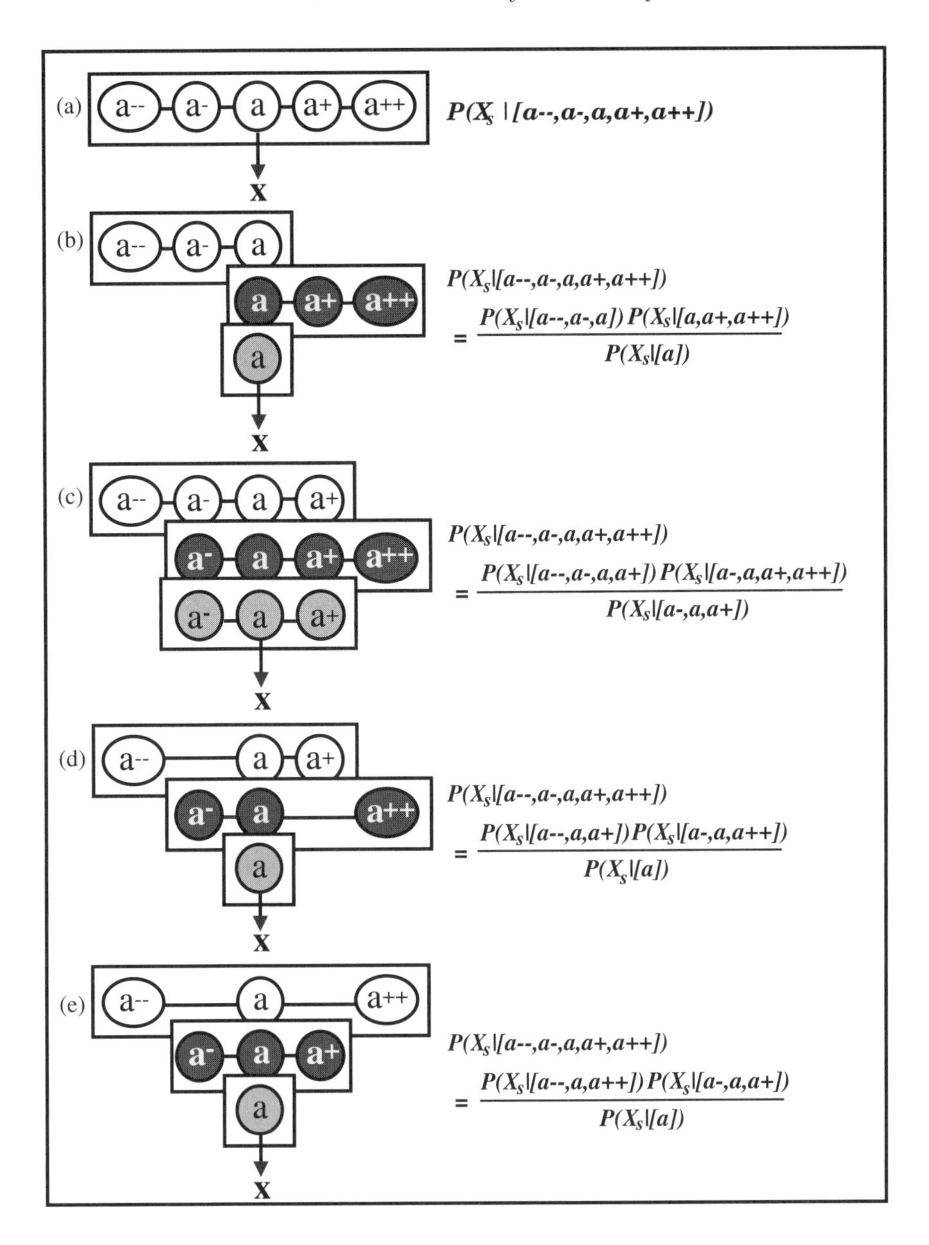

Fig. C.1. Bayesian pentaphone model composition. (a) is C5, the conventional pentaphone model, (b) is Bayesian C1L3R3, which is composed of the preceding/following triphone-context unit and center-monophone unit, (c) is Bayesian C3L4R4, which is composed of the preceding/following tetraphone-context unit and center-triphone-context unit, (d) is Bayesian C1Lsk3Rsk3, which is composed of the preceding/following skip-triphone-context unit and center-monophone unit, and (e) is Bayesian C1C3Csk3, which is composed of the center skip-triphone-context unit, center triphone-context unit and center-monophone unit.

D

Statistical Significance Testing

In this book, a statistical significance test was performed to determine whether the performance improvement of our proposed models is statistically significant in comparison with the baseline models. Various different types of significant test exist (Hays, 1988; Milton and Arnold, 1995). Here, we used one of the simplest significant tests called the "Sign test." A brief theory of general statistical hypothesis testing and the use of Sign test for ASR are described in Section D.1 and Section D.2, respectively.

D.1 Statistical Hypothesis Testing

A statistical hypothesis test is an inferential process, based on probability, which allows us to use sample data to evaluate a hypothesis about a population. The most basic approach to hypothesis testing consists of following steps:

1. State a research hypothesis

   The first step is to state a null and an alternative hypothesis which reference the population values:

   - Null hypothesis (H_0) represents the status quo or a claim of "no difference."
   - Alternative hypothesis (H_1) is the opposing hypothesis. That is, in our case, it is a claim of "there is a difference, and our proposed model significantly performed better" (upper-tail form).

2. Collect evidence (data sample)

   Evidence or data samples are collected from the population after the proposed approach has been applied.

3. Perform test statistic and obtain probability value (P-value)

   A test statistic is a numeric value used to decide whether to accept or reject H_0, by comparing the observed data samples to the population when H_0 is true. There are different types of test statistics depending on the hypothesis and the data being tested. For a normal distribution, the test statistic computes a z_{stat} value by calculating the difference between the observed sample $\bar{x}$ and the mean hypothesis value μ_0 as follows:

$$z_{stat} = \frac{\bar{x} - \mu_0}{\sigma_{\bar{x}}} \tag{D.1}$$

   where z_{stat} follows the $\mathcal{N}(0, 1)$ standardized distribution.

   The conditional probability of the observed sample given H_0 is true is called the P-value. It is the area under the curve beyond the z_{stat}.

$$P(Z \geq z_{stat} |\; H_0\; true) \tag{D.2}$$

   The smaller the P-value (the proportion of sample data that are consistent with the null hypothesis), the more likely that we will reject H_0.

4. Define decision rule and level of significance

   There is a chance that the hypothesis testing will yield inaccurate results, and rejecting the null hypothesis when H_0 is true is a serious mistake.

   The probability of rejecting H_0 when it is actually true is known as the significance level and is abbreviated with the symbol α. The chance of not committing this type of error is known as the confidence level, the probability of which equals $(1 - \alpha)$. Thus, to measures the reliability of the inference, α must be small. The commonly used levels of significance are 5%, 1% and 10%.

5. Make a conclusion

   Finally, the P-value is compared with probability threshold α (See Figure D.1), and a decision is made to either accept or reject H_0, as follows:

- Rule 1:

  If: Probability value ≤ significance level
  where:
  $$P(|Z| \geq z_{stat}) \leq \alpha$$
  Then: Reject the null hypothesis H_0.
  Conclusion: The research finding is statistically significant.

- Rule 2:

  If: Probability value > significance level
  where:
  $$P(|Z| \geq z_{stat}) > \alpha$$
  Then: Fail to reject the null hypothesis H_0.
  Conclusion: The research finding is not statistically significant.

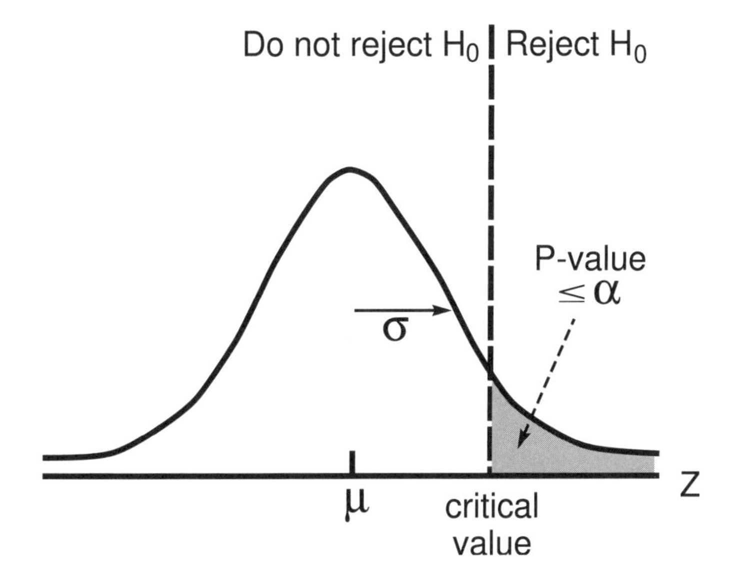

Fig. D.1. The distribution of population according to the null hypothesis (H_0 is true), with upper-tail of rejection region for $P \leq \alpha$.

D.2 The Use of the Sign Test for ASR

The Sign test is a nonparametric test that does not depend on a normal assumption for the population of differences (Larget, 2005). It is usually used for matched-pair data to discover if observed differences between two sets of data are significant.

The Sign test, first suggested for use in ASR benchmark test in (Pallett et al., 1992), is a test comparing WER on different speakers, conversation sides, or pre-specified subsets of a test set. It looks simply at which systems performs better. If there is systematic evidence of differences in a consistent direction, this may prove to be significant even if the magnitudes of the differences are small. In practice, we performed the statistical sign test to compare WER in each of the utterances of a test set.

Following the basic approach described in Section D.1, the hypothesis testing is defined as follows:

1. Research hypothesis

   Suppose there are two approaches, the baseline approach (A_0) and the proposed approach (A_1), then the hypothesis are

   H_0: $p \leq 0.5$
   H_1: $p > 0.5$

   where:

   p = the probability of performance improvement or success (A_1 is better than A_0)
   q = the proportion (probability) of failure (A_1 is not better than A_0)

2. Collect evidence (data sample)

   During recognition, both approaches (A_0 and A_1) output a sequence utterances with the word labels. The total number of word errors (including insertion, deletion, and substitution) in each utterance are described as

$$e_{0_i} = e_{0_1}, e_{0_2}, ..., e_{0_n}$$
$$e_{1_i} = e_{1_1}, e_{1_2}, ..., e_{1_n}$$

   where e_{0_i} is the total number of word errors in i^{th} utterance of A_0 output. The evidence for matched pair samples is obtained as:

- Calculate the differences of the paired observations (compare e_{0_1} with e_{1_1}, e_{0_2} with e_{1_2}, and so on).

- Discard the differences that are equal to zero, leaving N observations of pairs with nonzero differences.

- Count the sign of the differences, where:

  n^+ = the number of success ($e_{1_i} < e_{0_i}$)
  n^- = the number of failed ($e_{1_i} > e_{0_i}$)

3. Test statistic for the sign test

- For small samples

  The test statistic is based on binomial distribution $b(x; N, p)$, and P-value is computed as

$$P\text{-value} = P(X \geq n^+|N, p)$$

- For large samples

  The test statistic is based on a normal approximation (continuous) to the binomial distribution

$$\begin{aligned} z_{stat} &= \frac{\bar{x} - \mu_0}{\sigma_{\bar{x}}} = \frac{\bar{x}^* - 0.5N}{0.5\sqrt{N}} \\ &= \frac{(\bar{x} - 0.5) - 0.5N}{0.5\sqrt{N}} = \frac{n^+ - 0.5 - 0.5(n^+ + n^-)}{0.5\sqrt{N}} \\ &= \frac{|n^+ - n^-| - 1}{\sqrt{N}} \end{aligned} \tag{D.3}$$

  where

  $\mu_0 = Np = 0.5N$
  $\sigma_{\bar{x}} = \sqrt{Npq} = 0.5\sqrt{N}$
  $\bar{x}^*$ = the corrected for continuity adjustment of discrete value $\bar{x}$

  Then the P-value is computed as

$$P\text{-value} = P(Z \geq z_{stat})$$

4. Significance level

   $\alpha = 0.05$ has been used in all our experiments.

5. Decision:

   H_0 is rejected if P-value ≤ 0.05.

A more detailed description of significant hypothesis and the sign test can be found in (Hays, 1988; Milton and Arnold, 1995; Larget, 2005; Pallett et al., 1992).

References

Ahmed, N., Natarajan, T., Rao, K. R., 1974. Discrete cosine transform. IEEE Trans. Computers, 90–93.

Aubert, X., Dugast, C., Ney, H., Steinbiss, V., 1994. Large vocabulary continuous speech recognition of Wall Street Journal data. In: Proc. ICASSP. Adelaide, Australia, pp. 129–132.

Aull, A. M., Zue, V. W., 1985. Lexical stress determination and its application to large vocabulary speech recognition. In: Proc. ICASSP. Tampa, Finland, pp. 1549–1552.

Bahl, L., de Souza, P., Gopalakrishnan, P., Nahamoo, D., Picheny, M., 1991. Decision tree for phonological rules in continuous speech. In: Proc. ICASSP. Toronto, Canada, pp. 185–188.

Baker, J., 1975. The Dragon system, an overview. IEEE Trans. ASSP 23 (1), 24–29.

Bartels, C., Duh, K., Bilmes, J., Kirchoff, K., King, S., 2005. Genetic triangulation of graphical models for speech and language processing. In: Proc. EUROSPEECH. Lisbon, Portugal, pp. 3329–3332.

Baum, L., 1972. An inequality and associated maximization technique in statistical estimation for probabilistic functions of Markov processes. Inequalities 3, 1–8.

Beattie, V., Edmonson, S., Miller, D., Patel, Y., Talvola, G., 1995. An integrated multi-dialect speech recognition system with optional speaker adaptation. In: Proc. EUROSPEECH. Madrid, Spain, pp. 1123–1126.

Beinum, F. K.-V., 1980. Vowel contrast reduction: An acoustic and perceptual study of Dutch vowels in various speech conditions. Ph.D. thesis, Academic Amsterdam.

Bernardo, J., 2001. Bayesian statistic. In: UNESCO Encyclopedia of Life Support Systems (EOLSS). ftp://matheron.uv.es/pub /personal/bernardo/BayesStat.pdf.

Bernardo, J., Smith, A., 1994. Bayesian Theory. John Wiley & Sons Inc., Chichester, West Sussex, England.

Beyerlein, P., Aubert, X., Haeb-Umbach, R., Harris, M., Klakow, D., Wandemuth, A., Molau, S., Pitz, M., Sixtus, A., 1999. The Philips/RWTH system for transcription of broadcast news. In: Proc. DARPA Broadcast News Workshop. Virginia, USA, pp. 151–155.

Billi, R., 1982. Vector quantization and Markov models applied to speech recognition. IEEE, 574–577.

Bilmes, J., April 1998. A Gentle Tutorial of the EM Algorithm and its Application to Parameter Estimation for Gaussian Mixture and Hidden Markov Models. ICSI, U.C. Berkeley, USA.

Bilmes, J., 1999. Natural statistical models for automatic speech recognition. Ph.D. thesis, U.C. Berkeley, USA.

Bilmes, J., Zweig, G., Richardson, T., Filali, K., Livescu, K., Xu, P., Jackson, K., Brandman, Y., Sandness, E., Holtz, E., Torres, J., Byrne, B., 2001. Discriminatively structured graphical models for speech recognition. Tech. rep., John Hopkins University Baltimore, Baltimore, USA.

Booth, T., 1967. Sequential Machines and Automata Theory, 1st Edition. John Wiley and Sons, Inc., New York.

Bourlard, H., Morgan, N., 1994. Connectionist Speech Recognition : A Hybrid Approach. Kluwer Academic Publishers, Massachusetts, USA.

Bouselmi, G., Fohr, D., Illina, I., Haton, J. P., 2006. Multilingual non-native speech recognition using phonetic confusion-based acoustic model modification and graphemic constraints. In: Proc. ICSLP. Pittsburgh, USA, pp. 109–112.

Brenner, N., Rader, C., 1976. A new principle for fast Fourier transformation. IEEE Transaction on Acoustics, Speech and Signal Processing 24, 264–266.

Chen, B., 2004. Speech signal representations. Lecture Notes of National Taiwan Normal University, Taipei, Taiwan.

Chigier, B., Spitz, J., 1990. Are laboratory databases appropriate for training and testing telephone speech recognizers? In: Proc. ICSLP. Kobe, Japan, pp. 1017–1020.

Choukri, K., 2007. Importance of evaluation in language technology & beyond. European Language Resources Association (ELRA), http://www.ist-chorus.org/telechargement/1182958995.pdf.

Chow, Y., Dunham, M., Kimball, O., Krasner, M., Kubala, G., Makhoul, J., Roucos, S., Schwartz, R., 1987. BYBLOS: The BBN continuous speech recognition system. In: Proc. ICASSP. Dallas Texas, USA, pp. 89–92.

Daoudi, K., Fohr, D., Antoine, C., 2000. A new approach for multi-band speech recognition based on probabilistic graphical models. In: Proc. ICSLP. Beijing, China, pp. 329–332.

Davis, E., Selfridge, O., 1962. Eyes and ears for computers. Proc. IRE 50, 1093–1101.

Davis, K., Biddulph, R., Balashek, S., 1952. Automatic recognition of spoken digits. Journal of the Acoustical Society of America 24, 637–642.

DeGroot, M., 1970. Optimal Statistical Decisions. McGraw-Hill.

Dempster, A., Laird, N., Rubin, D., 1977. Maximum likelihood from incomplete data via the EM algorithm. Journal of Royal Statistics Society 39, 1–38.

Denes, P. B., Mathews, M. V., 1960. Spoken digit recognition using time-frequency pattern matching. Journal of the Acoustical Society of America 32, 1450–1455.

Deng, L., O'Shaughnessy, D., 2003. Speech Processing: A Dynamic and Optimization-Oriented Approach. Marcel Dekker, Inc., New York, USA.

Docio-Fernandez, L., Cardenal-Lopez, A., Garcia-Mateo, C., 2006. TC-STAR 2006 automatic speech recognition evaluation: The UVIGO system. In: Proc. TC-STAR Workshop on Speech-to-Speech Translation. Barcelona, Spain, pp. 145–150.

Doss, M. M., 2005. Using auxiliary sources of knowledge for automatic speech recognition. Ph.D. thesis, École Polytechnique Fédérale De Lausanne.

Doss, M. M., Stephenson, T. A., Bourlard, H., 2003. Using pitch frequency information in speech recognition. In: Proc. EUROSPEECH. Geneva, Switzerland, pp. 2525–2528.

Dupont, S., Luettin, J., 2000. Audio-visual speech modeling for continuous speech recognition. IEEE Trans. on Multimedia 2 (3), 141–151.

Ephraim, Y., Merhav, N., 2002. Hidden Markov processes. IEEE Trans. Inform. Theory 48, 1518–1569.

Ferguson, J., 1980. Hidden Markov Models for Speech. IDA, Princeton.

Fetter, P., 1998. Detection and transcription of OOV words. Ph.D. thesis, TU Berlin, Berlin, Germany.

Finke, M., Rogina, I., 1997. Wide-context acoustic modeling in read vs. spontaneous speech. In: Proc. ICASSP. Munich, Germany, pp. 1743–1746.

Fomey, G., 1973. The Viterbi algorithm. IEEE 61 (3), 268–278.

Fosler-Lussier, E., Greenberg, S., Morgan, N., 1999. Incorporating contextual phonetics into automatic speech recognition. In: Proc. ICPhs. San Francisco, USA, pp. 611–614.

Friedman, N., Goldszmidt, M., 1998. Learning Bayesian network from data. Tech. rep., SRI International, http://www/dsv.su.se/ijcai-99/tutorials/d3.html.

Fukada, T., Tokuda, K., Kobayashi, T., Imai, S., 1992. An adaptive algorithm for Mel-cepstral analysis of speech. In: Proc. ICASSP. pp. 137–140.

Fukunaga, K., 1990. Introduction to Statistical Pattern Recognition, 2nd Edition. Academic Press Inc., San Diego, CA, USA.

Furui, S., 1986. Speaker independent isolated word recognition using dynamic features of speech spectrum. IEEE Transactions Acoustics, Speech, and Signal Processing 34, 52–59.

Gales, M., Young, S., 1992. An improved approach to the hidden Markov model decomposition. In: Proc. ICASSP. San Francisco, CA, USA, pp. 233–236.

Gales, M., Young, S., 1995. A fast and flexible implementation of parallel model combination. In: Proc. ICASSP. Detroit, MI, USA, pp. 133–136.

Ganapathiraju, A., Hamaker, J., Picone, J., Ordowski, M., Doddington, G., 2001. Syllable-based large-vocabulary continuous speech recognition. IEEE Trans. on Speech and Audio Processing 9 (4), 358–366.

Garofolo, J., Lamel, L., Fisher, W., Fiscus, J., Pallett, D., Dahlgren, N., 1993. DARPA TIMIT acoustic-phonetic continuous speech corpus CD-ROM. Technical Report NISTIR 4930, NIST.

Gauvain, J., Lamel, L., Adda, G., Adda-Decker, M., 1994. The LIMSI continuous speech dictation system: Evaluation on the ARPA Wall Street Journal task. In: Proc. ICASSP. Adelaide, Australia, pp. 557–560.

Ghorshi, S., Vaseghi, S., Yan, Q., 2006. Comparative analysis of formants of British, American and Australian accents. In: Proc. ICSLP. Pittsburgh, Pennsylvania, USA, pp. 137–140.

Girardi, A., February 2001. Automatic acoustic modeling using tied-mixture successive state split algorithm. Ph.D. thesis, Nara Institute of Science and Technology (NAIST), Nara, Japan.

Godfrey, J., Holliman, E., McDaniel, J., 1992. SWITCHBOARD: Telephone speech corpus for research and development. In: Proc. ICASSP. San Francisco, California, USA, pp. 517–520.

Gold, B., Morgan, N., 1999. Speech and Audio Signal Processing: Processing and Perception of Speech and Music. John Wiley & Sons Inc., New York, USA.

Gong, Y., 1995. Speech recognition in noisy environments: A survey. Speech Communication 16 (3), 261–291.

Hansen, J., 1988. Analysis and compensation of stressed and noisy speech with application to robust automatic recognition. Ph.D. thesis, Georgia Institute of Technology, Georgia, USA.

Harris, F., January 1978. On the use of windows for harmonic analysis with the discrete Fourier transform. IEEE 66, 51–83.

Hays, W. L., 1988. Statistics. Holt, Rinehart and Winston Inc., Orlando, Florida, USA.

Heckerman, D., 1995. A tutorial on learning with Bayesian networks. in learning in graphical models. Tech. Rep. MSR-TR-95-06, Microsoft Research, Advanced Technology Division, Microsoft Corporation, Cambridge, MA, USA.

Heid, S., Hawkins, S., 2000. An acoustical study of long-domain /r/ and /l/ coarticulation. In: 5th Seminar on Speech Production: Model and Data. Kloster Seeon, Germany, pp. 77–80.

Henton, C., 1992. Acoustic variability in the vowels of female and male speakers. Journal of the Acoustical Society of America 91 (4), 2387.

Hermansky, H., 1990. Perceptual linear predictive (PLP) analysis of speech. The Journal of The Acoustical Society of America 87, 1738–1752.

Hirsch, H., Pearce, D., September 2000. The AURORA experimental framework for the performance evaluations of speech recognition systems under noisy conditions. In: ISCA ITRW ASR 2000 Automatic Speech Recognition: Challenges for the Next Millennium.

Holmes, J., Holmes, W., 2001. Speech Synthesis and Recognition. Taylor & Francis, London, UK.

Holmes, W., Huckvale, M., 1994. Why have HMMs been so successful for automatic speech recognition and how might they be improved? Speech, Hearing and Language 8, 207–219.

Hori, T., Noda, Y., Matsunaga, S., 2003. Improved phoneme-history-dependent search method for large-vocabulary continuous-speech recognition. IEICE Trans. Inf. & Syst. E86-D (6), 1059–1067.

Howell, P., Young, K., Sackin, S., 1992. Acoustical changes to speech in noisy and echoey environments. ETRW: Speech Processing in Adverse Conditions, 223–225.

Huang, C., Chang, E., Zhou, J., Lee, K. F., 2000. Accent modeling based on pronunciation dictionary adaptation for large vocabulary Mandarin speech recognition. In: Proc. ICSLP. Beijing, China, pp. 818–821.

Huang, C., Chen, T., Chang, E., 2004. Accent issue in large vocabulary continuous speech recognition. International Journal of Speech Technology 7 (2/3), 141–153.

Huang, C., Darwiche, A., 1994. Inference in belief networks: A procedural guide. International Journal of Approximate Reasoning 11, 1–158.

Huang, X., Acero, A., Hon, H.-W., 2001. Spoken Language Processing. Prentice Hall, New Jersey, USA.

Hughes, G. W., 1959. On the recognition of speech by machine. Sc.d dissertation, Dept. of Electrical Engineering, MIT, Cambridge, USA.

Hultzen, L. S., Allen, J. H. D., Miron, M. S., 1964. Tables of Transitional Frequencies of English Phonemes. IL: University of Illinois Press.

Iseli, M., Shue, Y. L., Alwan, A., 2006. Age- and gender-dependent analysis of voice source characteristics. In: Proc. ICASSP. Toulouse, France, pp. 389–392.

Jelinek, F., 1976. Continuous speech recognition using statistical methods. Proc. IEEE 64 (4), 532–556.

Jensen, F., 1998. An Introduction to Bayesian Networks. UCL Press, Aalborg University, Denmark.

Jensen, F., 2001. Bayesian Networks and Decision Graphs. Springer, Aalborg University, Denmark.

Jitsuhiro, T., 2005. Automatic model generation for speech recognition. Ph.D. thesis, Department of Information Processing, Graduate School of Information Science, Nara Institute of Technology.

Jitsuhiro, T., Matsui, T., Nakamura, S., 2004. Automatic generation of non-uniform HMM topologies based on the MDL criterion. IEICE Trans. Inf. & Syst. E87-D (8), 2121–2129.

Jordan, M., 1999. Learning in Graphical Models. MIT Press, Cambridge, MA, USA.

Juang, B., Rabiner, L., 2005. Automatic Speech Recognition: A Brief History of the Technology, 2nd Edition. Elsevier Encyclopedia of Language and Linguistics.

Junqua, J.-C., Haton, J.-P., 1996. Robustness in Automatic Speech Recognition, Fundamental and Applications. Kluwer Academic Publishers, Massachusetts, USA.

Jurafsky, D., 2007. Speech recognition and synthesis: Feature extraction and acoustic modeling. Lecture Notes of Stanford University, USA.

Kikui, G., Sumita, E., Takezawa, T., Yamamoto, S., 2003. Creating corpora for speech-to-speech translation. In: Proc. EUROSPEECH. pp. 381–384.

Kingsbury, B., 1998. Perceptually inspired signal processing strategies for robust speech recognition in reverberant environments. Ph.D. thesis, Dept. of EECS, University of California, Berkeley, USA.

Kiss, I., Leppanen, J., Sivadas, S., 2006. Nokia's system for TC-STAR EPPS English ASR evaluation task. In: Proc. TC-STAR Workshop on Speech-to-Speech Translation. Barcelona, Spain, pp. 129–132.

Kjaerulff, U., Madsen, A., 2005. Probabilistic networks - an introduction to Bayesian networks and influence diagrams. Tech. rep., Department of Computer Science, Aalborg University, Aalborg, Denmark.

Klatt, D., 1977. Review of the ARPA speech understanding project. Acoustical Society of America 62, 1345–1366.

Klatt, D., Klatt, L., 1990. Analysis, synthesis, and perception of voice quality variations among female and male talkers. Journal of the Acoustical Society of America 87 (2), 820–857.

Kubala, F., Bellegarda, J., Cohen, J., Pallet, D., Phillips, M., Rajasekaran, R., Richardson, F., Riley, M., Rosenfeld, R., Roth, B., Weintraub, M., 1994. The hub and spoke paradigm for CSR evaluation. In: Proc. ARPA Workshop on Spoken Language Technology. Princeton, NJ, USA, pp. 37–42.

Kuchera, H., Francis, W. N., 1967. Computational Analysis of Present-Day American English. Brown University Press, Providence, Rhode Island, USA.

Kuehner, B., Nolan, F., 1999. The origin of coarticulation. In: Hardcastle, W., Hawlett, N. (Eds.), Coarticulation: Theory, Data, Techniques. Cambridge University Press, Cambridge, UK, pp. 7–30.

Kumpf, K., King, R., 1997. Foreign speaker accent classification using phoneme-dependent accent discrimination models and comparisons with human perception benchmarks. In: Proc. EUROSPEECH. Rhodos, Greece, pp. 2323–2326.

Lamel, L., Gauvain, J., Adda, G., Barras, C., Bilinski, E., 2006. The LIMSI 2006 TC-STAR transcription systems. In: Proc. TC-STAR Workshop on Speech-to-Speech Translation. Barcelona, Spain, pp. 123–128.

Larget, B., 2005. Paired samples. Lecture Notes of Departments of Botany and Statistics, University of Wisconsin, Madison, USA.

Lea, W., 1986. Trends in speech recognition. Speech Science Publications, 3–18.

Lee, C., 2004. From knowledge-ignorant to knowledge-rich modeling: a new speech research paradigm for next generation automatic speech recognition. In: Proc. ICSLP. Jeju, Korea, pp. 109–112.

Lee, C., Rabiner, L., 1989. A frame synchronous network search algorithm for connected word recognition. IEEE Trans. ASSP 37 (11), 1649–1658.

Lee, K.-F., Hayamizu, S., Hon, H.-W., Huang, C., Swartz, J., R.Weide, 1990. Allophone clustering for continuous speech recognition. In: Proc. ICASSP. pp. 749–752.

Lee, S., Potamanos, A., S.Narayanan, 1999. Acoustic of children speech: Developmental changes of temporal and spectral parameters. Journal of the Acoustical Society of America 105 (3), 1455–1468.

Leggetter, C., Woodland, P., 1995. Maximum likelihood linear regression for speaker adaptation of continuous density hidden Markov models. Journal of Computer Speech and Language 9, 171–185.

Lesser, V., Fennell, R., Erman, L., Reddy, D., 1975. Organization of the HEARSAY II speech understanding system. IEEE Trans. ASSP 23 (1), 11–23.

Li, J., Tsao, Y., Lee, C.-H., 2005. A study on knowledge source integration for candidate rescoring in automatic speech recognition. In: Proc. ICASSP. Philadelphia, PA, USA, pp. 837–840.

Lippmann, R., 1997. Speech recognition by machines and humans. Speech Communication 22, 1–15.

Liu, Y., Pascale, F., 2006. Multi-accent Chinese speech recognition. In: Proc. ICSLP. pp. 133–136.

Ljolje, A., Hindle, D., Riley, M., Sproat, R., 2000. The AT&T LVCSR-2000 system. In: Speech Transcription Workshop. University of Maryland, USA.

Loof, J., Bisani, M., Gollan, C., Heigold, G., Hoffmeister, B., Plahl, C., Schlueter, R., Ney, H., 2006. The 2006 RWTH parliamentary speeches transcription system. In: Proc. TC-STAR Workshop on Speech-to-Speech Translation. Barcelona, Spain, pp. 133–138.

Lowerre, B., 1976. The HARPY speech recognition system. Tech. rep., Carnegie Mellon University, USA.

Maekinen, V., 2000. Front-end feature extraction with Mel-scaled cepstral coefficients. Tech. rep., Laboratory of Computational Engineering, Helsinki University of Technology, Helsinki, Finland.

Makhoul, J., Schwartz, R., 1994. State of the art in continuous speech recognition. In: Roe, D., Wilpon, J. (Eds.), Voice Communication Between Humans and Machines. National Academic Press, pp. 165–198.

Mariani, J., 1991. Knowledge-based approaches versus mathematical model based algorithms: the case of speech recognition. In: Proc. 30th Conference on Decision and Control. Brighton, USA, pp. 841–846.

Markel, J., Jr., A. G., 1976. Linear prediction of speech. Springer-Verlag.

Markov, K., Dang, J., Lizuka, Y., Nakamura, S., 2003. Hybrid HMM/BN ASR system integrating spectrum and articulatory features. In: Proc. EUROSPEECH. Geneva, Switzerland, pp. 965–968.

Markov, K., Nakamura, S., 2003. A hybrid HMM/BN acoustic model for automatic speech recognition. IEICE Trans. Inf. & Syst. E86-D (3), 438–445.

Markov, K., Nakamura, S., 2005. Modeling successive frame dependencies with hybrid HMM/BN acoustic model. In: Proc. ICASSP. Philadelphia, PA, USA, pp. 701–704.

Markov, K., Nakamura, S., 2006. Forward-backwards training of hybrid HMM/BN acoustic models. In: Proc. ICSLP. Pittsburgh, USA, pp. 621–624.

Martin, F., Shikano, K., Minami, Y., 1993. Recognition of noisy speech by composition of hidden Markov models. In: Proc. EUROSPEECH. Berlin, Germany, pp. 1031–1034.

Matsuda, S., Jitsuhiro, T., Markov, K., Nakamura, S., 2006. ATR parallel decoding speech recognition system robust to noise and speaking style. IEICE Trans. Inf. & Syst. E89-D (3), 989–997.

Messina, R., Jouvet, D., 2004. Context dependent long units for speech recognition. In: Proc. ICSLP. Jeju Island, Korea, pp. 645–648.

Meyn, S. P., Tweedie., R. L., 1993. Markov Chains and Stochastic Stability. Springer-Verlag, London.

Milton, J., Arnold, J., 1995. Introduction to probability and statistics: principles and applications for engineering and the computing sciences. McGraw Hill, Inc.

Ming, J., Boyle, P. O., Owens, M., Smith, F. J., November 1999. A Bayesian approach for building triphone models for continuous speech recognition. IEEE Trans. Speech and Audio Processing 7 (6), 678–684.

Ming, J., Smith, F. J., 1998. Improved phone recognition using Bayesian triphone models. In: Proc. ICASSP. Seattle, USA, pp. 409–412.

Mohri, M., Pereira, F., Riley, M., 2002. Weighted finite-state transducers in speech recognition. Computer Speech and Language 16, 69–88.

Morgan, N., Gold, B., 1999. Cepstrum analysis. Lecture Notes of College of Engineering Department of Electrical Engineering and Computer Sciences, University of California Berkeley, USA.

Morgan, N., Tajchman, G., Mirghafori, N., Konig, Y., Wooters, C., 1994. Scaling a hybrid HMM/MLP system for large vocabulary CSR. In: Proc. ARPA Spoken Language Technology Workshop. Princeton, NJ, USA, pp. 123–124.

Mostefa, D., Garcia, M., Hamon, O., Moreau, N., 2006. Tc-star evaluation report. Tech. Rep. FP6-506738, ELDA.

Murphy, K., May 2001. Introduction to graphical models. Tech. rep., University of British Columbia.

Nakamura, S., Markov, K., Nakaiwa, H., Kikui, G., Kawai, H., Jitsuhiro, T., Zhang, J., Yamamoto, H., Sumita, E., Yamamoto, S., March 2006. The ATR multilingual speech-to-speech translation system. IEEE Trans. Audio, Speech and Language Processing 14 (2), 365–376.

Neti, C., Potamianos, G., Luettin, J., Matthews, I., Glotin, H., Vergyri, D., Sison, J., Mashari, A., Zhou, J., 2000. Audio-visual speech recognition. Tech. rep., CSLP John Hopkins University, Baltimore, USA.

Neti, C., Roukos, S., 1997. Phone-context specific gender-dependent acoustic models for continuous speech recognition. In: Proc. ASRU. Santa Barbara, CA, USA, pp. 192–198.

Newell, A., Barnett, J., Forgie, J., Green, C., Klatt, D., Licklider, J. C. R., Munson, M., Reddy, R., Wood, W., 1971. Speech understanding systems: Final report of a study group. Tech. rep., Department of Computer Science, Carnegie Mellon University, Pittsburgh, USA.

Odell, J., 1995. The use of context in large vocabulary speech recognition. Ph.D. thesis, Cambridge University, Cambridge, UK.

O'Neill, P., Vaseghi, S., Doherty, B., Tan, W., McCourt, P., 1998. Multi-phone strings as subword units for speech recognition. In: Proc. ICSLP. Sydney, Australia, pp. 2523–2526.

Oppenheim, A., Schafer, R., 1975. Digital signal processing. Prentice-Hall.

Ostendorf, M., 2000. Incorporating linguistic theories of pronunciation variation into speech recognition models. Royal Society Philosophical Transactions 358 (1769), 1325–1338.

Ostendorf, M., Digalakis, V., 1991. The stochastic segment model for continuous speech recognition. In: Proc. 25th Asilomar Conference on Signals. Pacific Grove, CA, USA, pp. 964–968.

Pallett, D., 2003. A look at NIST's benchmark ASR tests: past, present, and future. In: Proc. ASRU. St. Thomas, Virgin Islands, USA, pp. 483–488, ©2003 IEEE.

Pallett, D., Fiscus, J., Fisher, W., Garofolo, J., Lund, B., Przybocki, M., 1994. 1993 benchmark tests for the ARPA spoken language program. In: Proc. ARPA Workshop on Spoken Language Technology. Princeton, NJ, USA, pp. 49–74.

Pallett, D., Fiscus, J., Garofolo, J., 1992. Resources management corpus: September 1992 test set benchmark results. In: Proc. ARPA Microelectronics Technology office Continuous Speech Recognition Workshop. Stanford, CA, USA.

Pallett, D., Fiscuss, J., Garofolo, J., Martin, A., Przybocki, M., 1999. 1998 broadcast news benchmark test results: English and non-English word error rate performance measures. In: Proc. DARPA Broadcast News Workshop. Virginia, USA, pp. 5–12.

Paul, D., Baker, J., 1992. The design for the Wall Street journal based CSR corpus. In: Proc. DARPA Workshop. Pacific Groove, California, USA, pp. 357–361.

Petrick, S. R., Willet, H. M., 1960. A method of voice communication with a digital computer. In: Proc. Eastern Joint Computer Conference. New York, USA, pp. 11–24.

Pfau, T., Beham, M., Reichl, W., Ruske, G., 1997. Creating large subword units for speech recognition. In: Proc. EUROSPEECH. Rhodos, Greece, pp. 1191–1194.

Picone, J. W., 1993. Signal modeling techniques in speech recognition. Proc. IEEE 81, 1215–1247.

Pierce, J., Kerlin, J., 1957. Reading rates and the information rate of a human channel. In: Bell System Technical Journal. No. 36. pp. 497–516.

Quast, H., 2001. Automatic recognition of nonverbal speech: An approach to model the perception of para- and extralinguistic vocal communication with neural networks. Ph.D. thesis, University of Goettingen, Germany.

Rabiner, L., 1989. A tutorial on hidden Markov models and selected applications in speech recognition. Proc. IEEE 77 (2), 257–286.

Rabiner, L., Juang, B.-H., 1993. Fundamentals of Speech Recognition. Prentice Hall, New Jersey, USA.

Ramabdharan, B., Siohan, O., Mangu, L., Zweig, G., Westphal, M., Schulz, H., Soneiro, A., 2006. The IBM 2006 speech transcription system for european parliamentary speeches. In: Proc. TC-STAR Workshop on Speech-to-Speech Translation. Barcelona, Spain, pp. 111–116.

Reichl, W., W.Chou, 1999. A unified approach of incorporating general features in decision-tree based acoustic modeling. In: Proc. ICASSP. Phoenix, Arizona, USA, pp. 573–576.

Riley, M., Pereira, F., Mohri, M., 1997. Transducer composition for context-dependent network expansion. In: Proc. EUROSPEECH. Rhodos, Greece, pp. 1427–1430.

Roberts, S., 2005. Lecture notes: Advanced probability theory. http://www.robots.ox.ac.uk/ sjrob/Teaching/AdvP/ap.html.

Robinson, T., Hochberg, M., Renals, S., 1994. Ipa: Improved phone modeling with recurrent neural networks. In: Proc. ICASSP. Adelaide, Australia, pp. 37–40.

Rosell, M., 2006. An introduction to frond-end processing and acoustic features for automatic speech recognition. Lecture Notes of School of Computer Science and Communication, KTH, Sweden.

Russel, S., Norvig, P., 1995. Artificial Intelligence, A Modern Approach. Prentice-Hall Inc.

Sagayama, S., 1989. Phoneme environment clustering for speech recognition. In: Proc. ICASSP. Glasgow, UK, pp. 397–400.

Sagayama, S., Yamaguchi, Y., Takahashi, S., Takahashi, J., 1997. Jacobian approach to fast acoustic model adaptation. In: Proc. ICASSP. Munich, Germany, pp. 835–838.

Sakti, S., 2005. A Tutorial Example of Using ATR Software Engine. ATR Spoken Language Communication Research Laboratories, Kyoto, Japan.

Sakti, S., Nakamura, S., Markov, K., 2005. Incorporating a Bayesian wide phonetic context model for acoustic rescoring. In: Proc. EUROSPEECH. Lisbon, Portugal, pp. 1629–1632.

Sakti, S., Nakamura, S., Markov, K., 2006. Improving acoustic model precision by incorporating a wide phonetic context based on a Bayesian framework. IEICE Trans. Inf. & Syst. E89-D (3), 946–953.

Scarborough, R., 2004. Coarticulation and the structure of the lexicon. PhD dissertation in Linguistics, University of California at Los Angeles (UCLA), USA.

Schuermann, J., 1996. Pattern Classification : A Unified View of Statistical and Neural Approach. John Wiley and Sons Inc., USA.

Schuster, M., Hori, T., 2005. Efficient generation of high-order context-dependent weighted finite state transducers for speech recognition. In: Proc. ICASSP. Philadelphia, PA, USA, pp. 201–204.

Scripture, E., 1902. The Elements of Experimental Phonetics. Charles Scribners Sons, New York, USA.

Seltzer, M., August 1999. Sphinx III signal processing front end specification. Tech. rep., CMU Speech Group, USA.

Seymore, K., McCallum, A., Rosenfeld, R., 1999. Learning hidden Markov model structure for information extraction. In: Proc. AAAI 99 Workshop on Machine Learning for Information Extraction. Orlando, Florida, USA, pp. 37–42.

Shachter, R., 1990. Evidence absorption and propagation through evidence reversals. Uncertainty in Artificial Intelligence, 173–190.

Shafran, I., Ostendorf, M., 2000. Use of higher level linguistic structure in acoustic modeling for speech recognition. In: Proc. ICASSP. Istanbul, Turkey, pp. 1021–1024.

Shinozaki, T., Furui, S., 2000. Dynamic Bayesian network-based acoustic models incorporating speaking rate effects. In: Proc. ICSLP. Beijing, China, pp. 329–332.

Siniscalchi, S., Gennaro, F., Andolina, S., Vitabile, S., Gentile, A., Sorbello, F., 2006. Embedded knowledge-based speech detectors for real-time recognition tasks. In: Proc. ICPP Workshops. Columbus, Ohio, USA.

Smith, E., Marian, S., Javier, M., 2001. Computer recognition of facial actions: A study of coarticulation effects. In: Proc. of the 8th Symposium of Neural Computation. California, USA.

Sorenson, H., 1980. Parameter Estimation: Principles and Problems. Marcel Dekker.

Stephenson, T., Mathew, M., Bourland, H., 2001. Modeling auxiliary information in Bayesian network based ASR. In: Proc. EUROSPEECH. Aalborg, Denmark, pp. 2765–2768.

Stern, P., Eskenazi, M., Memmi, D., 1986. An expert system for speech spectrogram reading. In: Proc. ICASSP. Tokyo, Japan, pp. 1193–1196.

Summers, W., Pisoni, D., Bernacki, R., Pedlow, R., Stokes, M., 1988. Effects of noise on speech production: Acoustic and perceptual analysis. Journal of the Acoustical Society of America 84 (3), 917–928.

Takami, J., Sagayama, S., 1992. A successive state splitting algorithm for efficient allophone modeling. In: Proc. ICASSP. San Francisco, CA, USA, pp. 573–576.

Takezawa, T., Sumita, E., Sugaya, F., Yamamoto, H., Yamamoto, S., 2002. Toward a broad-coverage bilingual corpus for speech translation of travel conversations in the real world. In: Proc. LREC. Las Palmas, Canary Islands, Spain, pp. 147–152.

Tolba, H., O'Shaughnessy, D., 2001. Speech recognition by intelligent machines. IEEE Canadian Review 38, 20–23.

Turn, R., 1974. The use of speech for man-computer communication. Tech. Rep. RAND Report-1386-ARPA, RAND Corp.

Valtchev, V., Odell, J., P.C.Woodland, Young, S., 1997. MMIE training of large vocabulary speech recognition systems. Speech Communication 22, 303–314.

Vergin, R., Farhat, A., O'Shaughnessy, D., 1996. Robust gender-dependent acoustic-phonetic modeling in continuous speech recognition based on a new automatic male/female classification. In: Proc. ICSLP. Philadelphia, PA, USA, pp. 1081–1084.

Viterbi, A., 1967. Error bounds for convolutional codes and an asymptotically optimum decoding algorithm. IEEE Trans. on Information Theory 13(2), 260–269.

Waibel, A., Lee, K.-F., 1990. Readings in Speech Recognition. Morgan Kaufman Publishers, San Mateo, CA, USA.

Wang, C., Seneff, S., 2001. Lexical stress modeling for improved speech recognition of spontaneous telephone speech in the Jupiter domain. In: Proc. EUROSPEECH. Aalborg, Denmark, pp. 2761–2764.

Wang, Z., Schultz, T., Waibel, A., 2003. Comparison of acoustic model adaptation techniques on non-native speech. In: Proc. ICASSP. Hong Kong, pp. 540–543.

Weintraub, M., Murveit, H., Cohen, M., Price, P., Bernstein, J., Baldwin, G., Bell, D., 1989. Linguistic constrain in hidden Markov model based speech recognition. In: Proc. ICASSP. Glasgow, UK, pp. 699–702.

Weintraub, M., Taussig, K., Hunicke-Smith, K., Snodgrass, A., 1996. Effect of speaking style on LVCSR performance. In: Proc. ICSLP. Philadelphia, PA, USA, pp. 16–19.

Werner, S., 2000. Fundamental of speech recognition. Student report, Department of Computer Science, Faculty of Electrical Engineering, University Duisburg, Duisburg, Germany.

West, P., 2000. Long distance coarticulatory effects of British English /l/ and /r/: and EMA, EPG and acoustic study. In: 5th Seminar on Speech Production: Model and Data. Kloster Seeon, Germany, pp. 105–108.

Wilpon, J., 1989. A study on effects of telephone transmission noise on speaker-independent recognition. In: Lea, W. (Ed.), Towards Robustness in Speech Recognition. Speech Science Publications, pp. 190–206.

Wilpon, J., Rabiner, L., 1994. Applications of voice-processing technology in telecommunications. In: Roe, D., Wilpon, J. (Eds.), Voice Communication Between Humans and Machines. National Academic Press, pp. 280–310.

Wolf, J., Woods, W., 1977. The HWIM speech understanding system. In: Proc. ICASSP. Hartford, Connecticut, USA, pp. 784–787.

Woodland, P., Odell, J., Valtchev, V., Young, S., 1994. Large vocabulary continuous speech recognition using HTK. In: Proc. ICASSP. Adelaide, Australia, pp. 125–128.

Young, S., Odell, J., Woodland, P., 1994. Tree-based state tying for high accuracy acoustic modeling. In: Proc. ARPA Workshop on Human Language Technology. pp. 307–312.

Zhao, J., Zhang, X., Ganapathiraju, A., Deshmukh, N., Picone, J., 1999. Decision tree-based state tying for acoustic modeling. Tutorial in Institute for Signal and Information Processing, Department of Electrical and Computer Engineering, Mississippi State University, USA.

Zue, V., 1985. The use of speech knowledge in speech recognition. IEEE Special issue on Man-Machine Communication 73 (11), 1602–1615.

Zue, V., Lamel, L., 1986. An expert spectrogram reader: A knowledge-based approach to speech recognition. In: Proc. ICASSP. Tokyo, Japan, pp. 1197–1200.

Zue, V., S.Seneff, Glass, J., 1990. Speech database development at MIT: TIMIT and beyond. Speech Communication 9 (4), 351–356.

Zweig, G., Russell, S., 1998. Probabilistic modeling with Bayesian networks for automatic speech recognition. In: Proc. ICSLP. Sydney, Australia, pp. 3010–3013.

Index

A priori probability, 50, 52
A-STAR, *see* Asian speech translation advanced research
A/D conversion, 38
Accents, 11
Acoustic model, 35, 43–49
  context-dependent model, 46
    following, 46
    preceding, 46
  continuous observation, 45
  decision tree clustering, 48
  discrete observation, 44
  parameter tying, 47
  phoneme-unit, 46
  syllable-unit, 46
  word-unit, 46
Acoustic phonetics, 4
Acoustic-phonetic knowledge, 13
Advanced Research Projects Agency (ARPA), 134
Advanced Telecommunication Research (ATR), 89, 104, 105, 112, 121, 128, 129
Algorithm complexity, 26–29
AM, *see* Acoustic model
AM score, 86
ARPA, *see* Advanced Research Projects Agency
Articulation, 13
Articulatory movements, 4
Articulatory pattern, 11
Asian speech translation advanced research (A-STAR), 150
ASR, *see* Automatic speech recognition
ATR, *see* Advanced Telecommunication Research
ATRASR tools, 153–161
AURORA, 145
Automatic speech recognition (ASR)
  acoustic model for, *see* Acoustic model
  approaches of
    corpus-based, 6–7
    knowledge-based, 4–6
  components of, 35
  definition of, 1
  difficulty, 7–10
  feature extraction for, *see* Feature extraction
  history, 2–4
  language model for, *see* Language model
  pattern recognition for, 35
  performance of, 7
  pronunciation lexicon for, *see* Pronunciation lexicon
  search algorithm for, *see* Search algorithm
Auxiliary function, 30–31
Auxiliary information, 13–14

Background noise, 11
Backward algorithm, 28
  backward probability, 28
  backward recurrence, 28
Bakis model, 25, 43
Bandwidth limitation, 11

Bandwiths, 40
Basic travel expression corpus (BTEC), 104, 105, 112, 121, 126, 128, 150
Baum-Welch algorithm, *see* Expectation-Maximization algorithm
Bayes's Rule, 58, 59, 163
Bayesian composition, 123, 126
Bayesian framework, 56
Bayesian models, 119, 124
Bayesian network (BN), 14, 17, 59–72
  direct inference, 71
  inference, 70
  joint PDF, 62, 70
  topology, 71, 74
Bayesian pentaphone, 119, 123–126, 131, 132, 136, 141, 166
Bayesian principle, 163
Bayesian triphone, 118, 119, 122, 125, 163
Bayesian wide-phonetic context model, 119, 121, 163–166
Benchmark ASR, 7
Bigram, 50
Binomial distribution, 173
BN, *see* Bayesian network
BTEC, *see* Basic travel expression corpus

C-STAR, *see* Consortium for Speech Translation Advanced Research
CCCC, *see* CSR Corpus Coordinating Committee
Cepstral mean subtraction (CMS), 42
Cepstrum analysis, 36
Chain rule, 58
Channel transmission, 11
Chord, 65
Chordal, 65
Chordless cycles, 65
Clique, 65
Clique potential, 67
Cluster potential, 67
CMS, *see* Cepstral mean subtraction
Coarticulation, 10, 12, 14, 46
Communication channel, 1
Conditional independence, 61
Conditional normalization, 57
Conditional probability, 57, 58, 60
Conditional probability distribution (CPD), 61
Conditional probability table (CPT), 61
Consonant, 10
Consortium for Speech Translation Advanced Research (C-STAR), 150
Continuous-speech recognition (CSR), 3, 148
Contour, 13
CORTES, 8
Covariance matrix, 45
CPD, *see* Conditional probability distribution
CPT, *see* Conditional probability table
Cross-word decoding, 12
CSR, *see* Continuous-speech recognition
CSR Corpus Coordinating Committee (CCCC), 148
Cycle, 65

DAG, *see* Directed acyclic graph
DARPA, *see* Defense Advanced Research Projects Agency
Data samples, 170
DBN, *see* Dynamic BN
DCT, *see* Discrete cosine transform
Defense Advanced Research Projects Agency (DARPA), 148
Deleted interpolation (DI), 86
DFT, *see* Discrete Fourier transform
DI, *see* Deleted interpolation
Digitization, *see* A/D conversion
Directed acyclic graph (DAG), 60
Discrete cosine transform (DCT), 40
Discrete Fourier transform (DFT), 37, 39–40
Distribued speech recognition (DSR), 95
DSR, *see* Distribued speech recognition
Dynamic BN (DBN), 14

EM, *see* Expectation-Maximization algorithm
Engine noise, 11
Enhancement approach, 85
  hard decision, 85
  no decision, 85
  soft decision, 85

EPPS, *see* European Parliament Plenary Sessions
European Parliament Plenary Sessions (EPPS), 8
Evidence, 170
Expectation-Maximization (EM) algorithm, 30–34
  E-step, 30, 32
  M-step, 30, 33

Fast Fourier transform (FFT), 40
Feature extraction, 35–42
  definition of, 36
Feature-based method, 14
Feature-based techniques, 14
FFT, *see* Fast Fourier transform
Filter bank, 36, 40
Filtering, 38
  low-pass, 38
  preemphasis, 39
Formant frequency, 1, 11
Formant peaks, 4
Forward algorithm, 27
  forward probability, 27
  forward recurrence, 27
Forward-backward probability, 31
Forward-backward recurrence, 31
Framing, 39
Fundamental frequency, 11

Gamma recursion, *see* Forward-backward recurrence
Gaussian densities, 68
Gaussian mixture model (GMM), 45
GDHMM, *see* Gender-dependent HMM
Gender-dependent HMM (GDHMM), 92
Gender-independent HMM (GIHMM), 92
GFIKS, *see* Graphical framework to incorporate knowledge sources
GIHMM, *see* Gender-independent HMM
Global conditional probability, 71
Global posterior probability, 52
Global property, 68
GMM, *see* Gaussian mixture model
Graph, 59
  arc reversal, 59
  arcs, 59
  arrow directions, 61
  connected, 65
  directed, 59, 67
  edges, 59, 65
  nodes, 59
  undirected, 59, 64
  vertices, 59, 65
Graph connection
  converging, 61, 62
  diverging, 61
  serial, 61
Graph theory, 56, 59
Graphical framework to incorporate knowledge sources (GFIKS), 16, 68
  at HMM phonetic-unit level, 83
  at HMM state level, 79
  for ASR, 79–137
  practical use of, 75–77
  procedure, 68–75
Graphical model, 56, 59–68

Hamming window, 39
Handset, 11
Hidden Markov model (HMM)
  backward algorithm for, *see* Backward algorithm
  decoding of, *see* Viterbi algorithm
  definition of, 23
  elements of, 23
  evaluation of, 25–28
  forward algorithm for, *see* Forward algorithm
  general form of, 23–25
  principle cases of, 25–34
  state distribution, 43
  state of, 23
  state sequence, 23
  state transition, 43
  training of, *see* Expectation-Maximization algorithm
Hypothesis testing, 169–171
  alternative hypothesis, 169
  mean value, 170
  null hypothesis, 169

IDFT, *see* Inverse discrete Fourier transform, *see* Inverse discrete Fourier transform
Intra-word decoding, 13
Inverse discrete Fourier transform (IDFT), 37, 40
Isolated-digit recognition, 2
Isolated-word recognition, 2

Jacobian matrix, 14
Joint probability function, 66
Junction graph, 65
Junction tree, 65
Junction tree algorithm, 63–68, 73
Junction Tree Decomposition, 72
Junction tree inference, 66, 75

Knowledge incorporation, 12
  contextual, 12
  environmental, 14
  speaker, 13
Knowledge sources
  contextual information, 10
  environmental information, 11
  speaker information, 10

Lagrange multipliers, 33
Language model, 35
Language model (LM), 50–51
  back-off mechanism, 51
Large-vocabulary continuous-speech recognition (LVCSR), 3, 5, 15
Lexical stress, 11, 13
Likelihood, 52
Linear prediction coefficients (LPC), 36
Linguistics, 4
LM, *see* Language model (LM)
LM score, 86
Logarithmic, 40
Longitudinal speaker dependent (LSD), 149
LPC, *see* Linear prediction coefficients
LSD, *see* Longitudinal speaker dependent
LVCSR, *see* Large-vocabulary continuous-speech recognition (LVCSR)

Machine translation (MT), 151
Machine translation aided dialogue (MAD), 150
MAD, *see* Machine translation aided dialogue
MAP, *see* Maximum a posteriori
Markov assumption, 23
Markov chain, 22–23, 43
  definition of, 22
  deterministic event, 22
Maximum a posteriori (MAP), 51
Maximum likelihood (ML), 49, 83, 89, 95, 104, 112
Maximum likelihood linear regression (MLLR), 14
MDL, *see* Minimum description length
Mel scale, 40
Mel-frequency cepstral coefficients (MFCC), 36, 38
MFCC, *see* Mel-frequency cepstral coefficients
Minimum description length (MDL), 49
ML, *see* Maximum likelihood
MLLR, *see* Maximum likelihood linear regression
Model-based method, 14
Modulation-filtered spectrograms (MSG), 36
Monophone, 12
Moral graph, 64
Moralization, 64
MSG, *see* Modulation-filtered spectrograms
MT, *see* Machine translation
Multi-condition training, 14
Multi-speaker ASR, 3
Multiaccent ASR, 3
Multilingual ASR, 3
Multiphone, 12
Multiple AM, 14

N-gram, 50
National Institute of Standards and Technology (NIST), 7
Neural networks, 13
NIST, *see* National Institute of Standards and Technology
Noise robustness, 14
Noise voice composition (NOVO), 14
Non-verbalized punctuation, 149

Normal distribution, 170
NOVO, *see* Noise voice composition
Nyquist criterion, 38

Observation
  density, 16, 19, 43–46
  instances, 26
  probability, 23
  probability distribution, 24
  probability function, 25
  sequence, 6, 25, 30
  symbols, 24
  variable, 26
Observation feature vectors, 35, 43
Offline recognition, 53
Online recognition, 53
Oval nodes, 72

P-value, *see* Probability value
Parallel model combination (PMC), 14
Partial observation, 29
Pattern
  definition of, 19
Pattern classification
  decision space, 20
  definition of, 20
  fundamental notions, 20
  measurement space, 20
Pattern recognition
  definition of, 19
  overview of, 19–22
Pentaphone, 12
Perceptual linear prediction (PLP), 36
Phoneme segments, 10
Phonetic unit, 12
Phonotactics, 4
Physiological state, 11
Pitch frequency, 13
PLP, *see* Perceptual linear prediction
PMC, *see* Parallel model combination
Prior distribution, 58
Priors, *see* A priori probability
Probabilistic networks, *see* Bayesian network
Probability
  axioms, 56–57
  calculus, 56
  emission, 43
  likelihood, 58
  posterior probability, 58
Probability density function (PDF), 68
Probability product, 58
Probability theory, 56–58
Probability value (*P*-value), 170, 173
Pronunciation lexicon, 35, 49
  lexical tree-based search, 49
  lexicon tree, 49
Psychological state, 11

Quasi-stationary, 39
Quinphone, 12

Recognition
  definition of, 19
Rescoring, 86
  N-best, 86
  procedure, 86
Rex, 2

S2ST, *see* Speech-to-speech translation
SD, *see* Speaker dependent
Search algorithm, 35, 51–53
  Viterbi, *see* Viterbi algorithm
Sennheiser microphone, 149
Separator potential, 67
SI, *see* Speaker independent
Sign test, 172–174
Signal-to-noise ratio (SNR), 76, 94–97
Significance level, 170, 174
Single-word recognition, 2
Small-vocabulary ASR, 2
SNR, *see* Signal-to-noise ratio
Sound source, 1, 2
Source-filter model, 36
Sources variability, 10
Speaker dependent (SD), 149
Speaker independent (SI), 149
Speaking rate, 11, 14
Spectral analysis, 39
Spectral shaping, 38
Spectrogram
  definition of, 4
Spectrogram reading, 5
Speech enhancement, 14
Speech unit, 12
Speech-to-speech translation (S2ST), 150
Square nodes, 72

SSS, *see* Successive state splitting
Stochastic rule, 6
Stochastic variables, 6
Sub-band correlation, 14
Subgraphs, 65, 67
Successive state splitting (SSS), 48, 49
SWB, *see* SWITCHBOARD
SWITCHBOARD (SWB), 12
Syllable unit, 12

TC-STAR, *see* Technology and Corpora for Speech to Speech Translation(TC-STAR)
Technology and Corpora for Speech to Speech Translation(TC-STAR), 8
Telephone speech, 11
Tetraphone, 12
TIDigits, 145
TIMIT, 146
Transition probabilities, 24
Trellis diagram, 26, 29
Triangular filter bank, 40
Triangulated graph, 65
Triangulation, 65
Trigram, 50
Triphone, 12
Triplets of nodes, 65

Unigram, 51

Vector quantization (VQ), 44
Verbalized punctuation, 149
Viterbi algorithm, 29
  best path probability, 29
Viterbi path, 45
VOA, *see* Voice of America
Vocal tract, 4, 11
Voice of America (VOA), 8
Voice quality, 11
Vowel, 10
VQ, *see* Vector quantization

Wall Street Journal (WSJ), 89, 90, 92, 104, 121, 148
Weighted Finite State Transducers (WFST), 13
WER, *see* Word error rate
WFST, *see* Weighted Finite State Transducers
Windowing, 39
Word error rate (WER), 91–93, 108, 114, 125, 172
Word position, 13
Word unit, 12
WSJ, *see* Wall Street Journal

Continued from page ii

Informatics in Control, Automation and Robotics: Selected Papers from the International Conference on Informatics in Control, Automation and Robotics 2007
Filipe, J.B.; Ferrier, Jean-Louis; Andrade-Cetto, Juan (Eds.)
978-3-540-85639-9

Digital Terrestrial Broadcasting Networks
Beutler, Roland
ISBN 978-0-387-09634-6

Logic Synthesis for Compositional Microprogram Control Units
Barkalov, Alexander, Titarenko, Larysa
ISBN: 978-3-540-69283-6

Sensors: Advancements in Modeling, Design Issues, Fabrication and Practical Applications
Mukhopadhyay, Subhas Chandra; Huang, Yueh-Min (Eds.)
ISBN: 978-3-540-69030-6

Smart Sensors and Sensing Technology
Mukhopadhyay, Subhas Chandra; Sen Gupta, Gourab (Eds.)
ISBN: 978-3-540-79589-6

Basic Principles of Fresnel Antenna Arrays
Minin, Igor V., Minin, Oleg V.
ISBN: 978-3-540-79558-2

Fundamental Numerical Methods for Electrical Engineering
Rosloniec, Stanislaw
ISBN: 978-3-540-79518-6

RFID Security and Privacy: Concepts, Protocols, and Architectures
Henrici, Dirk
ISBN: 978-3-540-79075-4

Advances in Mobile and Wireless Communications: Views of the 16th IST Mobile and Wireless Communication Summit
Frigyes, István; Bito, Janos; Bakki, Péter (Eds.)
ISBN: 978-3-540-79040-2

Informatics in Control Automation and Robotics: Selected Papers from the International Conference on Informatics in Control Automation and Robotics 2006
Andrade Cetto, J.; Ferrier, J.-L.; Pereira, J.M.C.D.; Filipe, J. (Eds.)
ISBN: 978-3-540-79141-6

Bandwidth Extension of Speech Signals
Iser, Bernd, Minker, Wolfgang, Schmidt, Gerhard
ISBN: 978-0-387-68898-5

Proceedings of Light-Activated Tissue Regeneration and Therapy Conference
Waynant, Ronald; Tata, Darrell B. (Eds.)
ISBN: 978-0-387-71808-8

Printed in the United States of America